The Algorithmic Ocean

How Tiny Organisms and Currents Shape the Planet

Team Trex

Published by NobleTrex Press

For permissions and other inquiries, write to:
P.O. Box 3132, Framingham, MA 01701,
USA

Contents

Introduction

Consider, for a moment, the breath you just took. It is an act so reflexive, so woven into the background rhythm of existence, that we rarely pause to trace its origin. You might vaguely recall a grade-school biology lesson about rainforests, the "lungs of the planet," and imagine a tree in the Amazon filtering the air. But statistically speaking, that breath-or at least every second one you take-did not come from a tree. It did not come from grass, or a fern, or a moss.

It came from a ghost.

It was crafted by a single-celled speck of life floating in the middle of the Pacific Ocean, thousands of miles from the nearest shore. This organism, perhaps a *Prochlorococcus*, is so small that millions of them could fit into a single drop of water. It is invisible to the naked eye, a ghost in the machine of the world. Yet, under the influence of the sun, it cracked open a water molecule, kept the hydrogen to build its tiny body, and discarded the oxygen as waste. That waste product, carried by the trade winds and dispersed into the stratosphere, is the fuel for your blood.

But here is the catch: that tiny organism is not swimming aimlessly. It is being driven. It is caught in a massive, swirling fluid engine that operates with the precision of a computer code. The ocean is not merely a tub of water; it is a computational system. It is a place where physics acts as the hardware and biology as the software. The rotation of the Earth, the drag of the wind, and the pull of the moon create a complex set of instructions-an algorithm-that dictates where life can exist, how the planet cools itself, and ultimately, whether human civilization can thrive.

This book is the story of that code. It is an exploration of the "Algorithmic Ocean," a planetary processor where tiny organisms and massive currents conspire to shape the world we walk upon.

To understand this system, we must abandon the intuition that the ocean is chaotic. To the casual observer standing on a beach, the waves look messy, crashing with unpredictable fury. But if you were to zoom out-way out, to a satellite's vantage point-you would see the order. You would see that the ocean moves in vast, sluggish gyres, turning clockwise in the north and counter-clockwise in the south, obeying laws of motion that are absolute.

This brings us to the "hardware" of our planetary machine. The ocean is a fluid in motion on a spinning sphere, and that spin changes everything. Early in the 20th century, the Swedish oceanographer Vagn Walfrid Ekman noticed something peculiar while observing icebergs in the Arctic: they didn't drift in the direction of the wind, but at an angle to it. He realized that the Earth's rotation applies a subtle force, twisting the water layers beneath the surface like a spiral staircase. This phenomenon, now known as Ekman transport, is the first line of code in our oceanic algorithm. It pushes water away from coastlines and away from the equator, forcing deep, cold water to rise and take its place.

This upward surge, or upwelling, is the power switch. The deep ocean is a warehouse of nutrients-nitrogen, phosphorus, iron-that have sunk into the abyss over centuries. When the physics of the spinning Earth drags this chemical treasure chest back to the sunlit surface, the "software" begins to run.

The result is the most massive explosion of life on Earth. In the span of a few days, trillions of phytoplankton bloom, turning the blue ocean into swirls of green, turquoise, and milk-white that are visible from the International Space Station. These are not just plants; they are solar panels. They harvest the energy of our star with an efficiency that human engineers envy. As we will see, scientists like Harald Sverdrup figured out that these blooms are timed perfectly to the depth of the mixing water, a delicate balance between being cooked by the sun and starved in the dark.

But why does this matter to you, beyond the oxygen in your lungs? Because this biological machinery is the primary reason Earth is not Venus.

We live on a planet that should be much hotter than it is. We have pumped enough carbon dioxide into the atmosphere since the Industrial Revolution to cook ourselves, yet the temperature has risen only a fraction of what the raw physics suggests. We have been bailed out by the ocean. The interface between the air and the waves is a frantic marketplace where gases are constantly swapped. The cold, dense waters of the North Atlantic act as a drain, sucking carbon dioxide out of the sky and sinking it to the bottom of the world.

This is the "Biological Pump," a mechanism as vital to our survival as the beating of our own hearts. Every time a plankton cell dies, or is eaten and excreted, it falls. This "marine snow"-a blizzard of dead cells, fecal pellets, and dust-drifts down through the twilight zone, carrying carbon away from the atmosphere for centuries, perhaps millennia. It is a cooling system of epic proportions, maintained by creatures that have no brain, no eyes, and no intent, yet collectively function as a planetary thermostat.

However, the ocean is not a peaceful garden. It is a battlefield. As we delve into the microscopic world, we find a "Hunger Games" scenario played out on a global scale. The phytoplankton are hunted by zooplankton, tiny animals that engage in the largest migration on Earth every single night, rising from the depths to graze under the cover of darkness and retreating at dawn to avoid predators. And they, in turn, are stalked by the most abundant biological entities in the ocean: viruses.

It is a staggering thought that in every milliliter of seawater, there are millions of viruses. These "viral code breakers" are essential to the system. They infect bacteria and algae, bursting them open and spilling their nutrients back into the water, short-circuiting the food web and ensuring that life at the surface doesn't run out of fuel. It is a ruthless loop of regeneration, a recycling program that wastes nothing.

As we journey through these pages, we will uncover the chemical ratios that govern this life. In the 1930s, Alfred Redfield made a discovery that haunts oceanography to this day. He found that no matter where you look in the deep ocean, the ratio of carbon, nitrogen, and phosphorus in the water is almost identical to the ratio found in the plankton bodies themselves-106 atoms of carbon for every 16 of nitrogen and 1 of phosphorus. This "Redfield Ratio" suggests a profound truth: the life in the sea and the chemistry of the sea are one and the same. The organisms have tuned the ocean to their needs, and the ocean has shaped the organisms in return.

Yet, the algorithm is glitching.

The ocean is a memory bank. It has a high "thermal inertia," meaning it takes a long time to heat up, but once it does, it holds onto that heat for centuries. We are currently loading the system with more energy than it has handled in millions of years. The result is not just warmer water; it is a rewriting of the code.

We are seeing the chemistry of the seawater change, a process known as acidification. As the ocean absorbs our excess carbon dioxide, its pH drops. This might sound like a minor adjustment, a line of code tweaked, but for the organisms that build shells out of calcium carbonate-from the microscopic pteropods at the base of the food web to the massive coral reefs-it is catastrophe. Their very skeletons are becoming harder to build, dissolving in the corrosive soup we have created.

Furthermore, the physical engine is slowing down. The great "Conveyor Belt" of currents, driven by the sinking of cold, salty water at the poles, distributes heat around the globe. It is the reason Europe is mild and the tropics are habitable. But as ice melts and freshens the polar seas, that sinking motion weakens. We face the paradox of a warming world potentially triggering a stagnation of the currents that sustain us.

This book is not a eulogy for the ocean, nor is it a blind celebration. It is an attempt to see the machine for what it is. We are moving past the era of wooden ships and message-in-a-bottle science. Today, we monitor the ocean's pulse with a "robotic fleet"-thousands of autonomous Argo floats that drift through the abyss, waking up every ten days to profile the water column and beam data to satellites. We are listening to the planet's heartbeat in real-time.

To understand the Algorithmic Ocean is to understand our own future. We are not separate from this system; we are a new variable introduced into an equation that has run smoothly for eons. The tiny organisms and the swirling currents do not care about political borders or economic quarters. They obey only the laws of fluid dynamics and stoichiometry.

So, let us dive in. We will start not with the biology, but with the motion-the spin, the wind, and the spiral. We must first understand the engine before we can meet the crew. The water is waiting, and the code is running.

Chapter 1

The Spinning Engine

The ocean is often imagined as a passive bowl, but it is actually a restless machine powered by the friction of wind and the rotation of the Earth. This chapter peels back the surface to reveal the counter-intuitive physics-spirals, vertical pumps, and chaotic eddies-that construct the dynamic architecture necessary for life.

1.1 The Spiral Staircase

When Fridtjof Nansen let his ship *Fram* freeze into the Arctic pack ice in the 1890s, he expected to be carried where the wind and the large-scale drift of the ice took him. The surprise was that the ice did not simply slide downwind. Day after day, the pack seemed to move at a stubborn angle to the breeze, as if an invisible hand were pushing sideways. In a landscape where a horizon can be nothing but pale sky and paler ice, a persistent sideways drift is not a small detail; it is a navigational fact. Nansen recorded it carefully. The puzzle then migrated south, out of the polar night and into the well-lit world of theory: what could make wind-driven motion turn?

The answer that emerged, in Vagn Walfrid Ekman's work (1905), is one of the ocean's quietest tricks. It does not require towering waves or dramatic storms, only three ingredients that are almost always present: a push at the surface (wind), a preference for turning rather than going straight (Earth's rotation), and a way for the surface's motion to tug on the water beneath (friction, mostly in its turbulent, churning form). Put together, they build a structure that is invisible from above but decisive for climate, for nutrients, and for tiny organisms trying to remain in the light: a spiralling stack of currents that changes direction with depth.

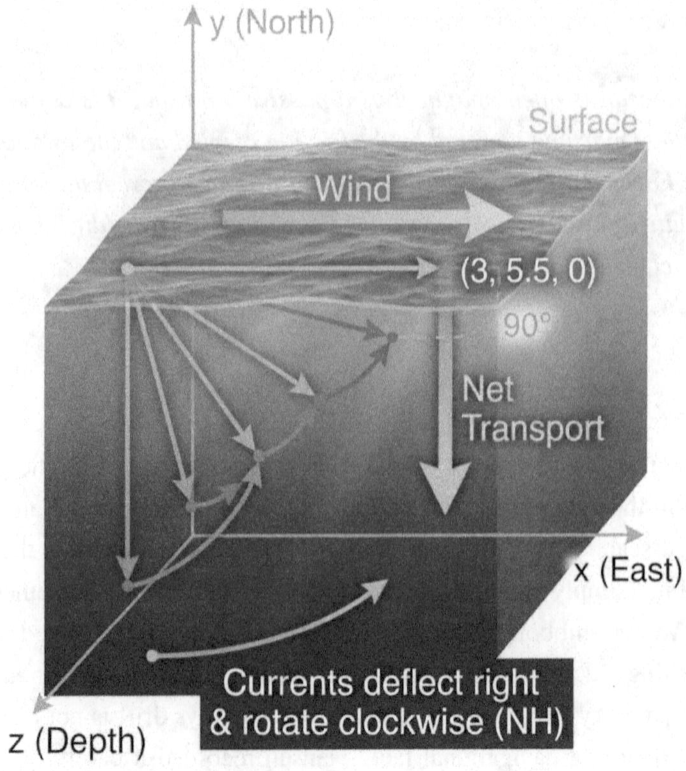

Imagine standing on a pier on a windy day, watching the surface of a bay. The wind roughens it, and you can almost see the skin of water being dragged along. That is not an illusion: air moving over water exerts a shear stress, a kind of continuous rubbing that transfers momentum into the ocean. But the ocean is not a rigid plate; it is layered and slippery. The surface moves most, the water just beneath is pulled along, and so on, with the influence fading as you go down. If Earth did not rotate, the story would be simple: the current would point roughly with the wind and weaken with depth, like a smudged fingerprint.

Rotation changes the rules because moving water on a rotating planet experiences an apparent sideways deflection. In the Northern Hemisphere, motion is turned to the right of its direction of travel; in the Southern Hemisphere, to the left. The effect is not a force you can bottle, but it behaves like one in the equations and in experience. It is the same geometric constraint that makes large weather systems spin and that makes long-range artillery and aircraft navigation quietly care about latitude. For ocean surface layers, it means that once the wind sets the water moving, the planet insists that the motion be nudged sideways.

That alone would be merely annoying: a current that veers until the sideways deflection balances the wind's push. But the ocean has depth, and the surface is not the only layer feeling a tug. Ekman's crucial idea was to treat the upper ocean like a stack of thin sheets, each one rubbing against the next. The wind pushes the top sheet. Earth's rotation turns that sheet's motion sideways. Friction transmits some of that motion downward, but weaker, because energy is lost to turbulent mixing. The second sheet therefore moves more slowly. Because the sideways turning depends on how fast the water is moving, the balance of forces in the second sheet is slightly different from the first. It ends up pointing in a different direction. Repeat this down the stack and you get a slow, graceful rotation with depth: the Ekman spiral.

The spiral is easiest to picture as a staircase whose steps are currents. At the very top, water runs at an angle to the wind rather than directly with it. A little deeper, the flow is weaker and turned a little more.

Deeper still, weaker again and turned further. In the idealized version drawn in textbooks, the current keeps rotating until it is almost opposite the surface flow, though in reality the ocean's turbulence, stratification, and changing winds often interrupt the clean geometry. What survives the messiness is the practical consequence that Nansen indirectly encountered: add up the motion of all the layers in the wind-driven boundary region, and the net transport is approximately perpendicular to the wind, to the right in the Northern Hemisphere and to the left in the Southern.

This is a strange conclusion if one's intuition is built on pushing solid objects. Push a crate, and it moves where you push it. Push the sea, and a good portion of its response is sideways. The ocean seems to be evasive, as if it refuses to be bullied directly. But the evasion is consistent. The sideways drift that bothered Nansen becomes predictable once the planet's spin is admitted as a constant background actor.

There is a second twist that makes the spiral more than a curiosity: the *net* transport is often the part that matters most for the living ocean. A surface current angled by, say, 20° or 30° may not sound life-changing. A net flow at 90° to the wind, extended over many kilometres and sustained for days, is another matter. It can shift surface waters away from a coastline, or pile them up against one. It can cause the sea surface to tilt slightly, setting up pressure gradients that feed larger currents. It can also reshape the thickness of the surface layer in which microscopic plants, the phytoplankton, can live.

To see how the spiral arises without drowning in mathematics, it helps to follow a parcel of water as if it had a modest biography. It begins at the surface, newly accelerated by the wind. In the Northern Hemisphere, as it starts to move, rotation turns it to the right. If the wind continues, the parcel does not simply curve forever, because friction and pressure adjustments matter too, but its direction is nudged away from the wind line. Now that surface parcel is sliding past the water just below. That shear produces turbulence: little whirls and eddies that act like a thickened kind of friction, transmitting momentum downward. The water beneath begins to move as well, but it starts later, so it remains slower. Rotation turns this slower motion too, but because the balance between turning and friction differs, it points at a different angle. Layer by layer, the ocean builds a rotating profile, like a corkscrew made of arrows.

What fixes the direction of the corkscrew is not the wind itself but the sign of rotation. In the Northern Hemisphere, the turning with depth is typically clockwise when you look downward, and the net transport is to the right of the wind. In the Southern Hemisphere it is reversed. This is one reason the world's coastal upwelling systems, despite sharing the same physics, have opposite wind directions on opposite sides of the equator. The geometry of rotation rewrites geography.

The depth over which the spiral matters is not a universal constant. One often hears a rule of thumb of around 100 m for an Ekman layer, but the real ocean is less obliging. In calm conditions with strong stratification (a stable layering where light, warm water floats over heavier, colder water), the wind's influence may be trapped in a shallower skin. During storms, with breaking waves and vigorous mixing, the wind's momentum can be hammered deeper. Sea ice can change the coupling between air and water, altering both the stress and the turbulence beneath. Even the daily heating and cooling cycle can matter: sunlight warms the surface and strengthens stratification by day; nighttime cooling and convection can deepen mixing. The spiral, then, is not a rigid staircase carved into the sea but a flexible one, whose steps shift with weather and season.

Yet even a flexible staircase can carry weight. Consider a concrete example: a summer wind blowing along the west coast of North

America. Locally, the wind often runs parallel to the shoreline. If the wind blows southward along a north—south coast in the Northern Hemisphere, the Ekman transport is to the right of the wind, meaning offshore, away from land. Offshore transport at the surface demands compensation: water must rise from below to replace what is moved outward, because the coastline blocks replacement from the landward side. This upwelling draws cold, nutrient-rich water into the sunlit layer, fertilizing blooms of phytoplankton that can, under the right conditions, feed entire food webs from zooplankton to fish to seabirds. The dramatic part is biological; the quiet engine underneath is mechanical, and it begins with that sideways transport implied by the spiral.

Even far from coasts, the same sideways logic can be felt. If winds over a broad region change in strength from place to place, the Ekman transport changes too. Where surface water converges, it is pushed downward; where it diverges, water rises. These vertical motions are often small in speed—centimetres per day can be significant—but they matter because the ocean is strongly layered in chemistry and temperature. A small vertical displacement can move water across steep gradients in nutrients, oxygen, or carbon chemistry. The spiral is thus connected, by simple mass conservation, to a vertical breathing of the ocean.

The historical fascination of Ekman's work is that it took an observation that sounded like a sailor's superstition—ice drifting askew to the wind—and gave it a quantitative spine. Ekman did not need satellites or autonomous floats; he needed the willingness to believe that the ocean's surface is coupled to its interior by turbulence, and that Earth's rotation, usually treated as a backdrop, is in fact an active constraint on everyday motion. In doing so, he helped make the ocean legible as a rotating fluid rather than a passive basin.

It is worth admitting how idealized the clean spiral can be. The ocean's surface rarely experiences a steady wind long enough to settle into a textbook Ekman balance. Gusts come and go; storms pass; waves break; currents already present can shear and twist the response. Turbulence is not a simple viscosity; it changes with stratification and with the very shear it is meant to transmit. Moreover, in much of the upper ocean, the flow is influenced by density variations—lighter water over heavier water—so the direction and strength of currents can also be steered by pressure gradients associated with sloping density surfaces. Real profiles measured by instruments can look ragged compared with the smooth corkscrew of theory.

But the ocean often forgives our imperfect pictures if we ask the right question. The most robust lesson is not that currents at each metre of depth form a perfect spiral, but that the *integrated* response of the wind-driven surface layer has a sideways character. For many practical purposes—from predicting where surface drifters or oil slicks may go, to understanding why certain coastlines are biologically rich, to interpreting the seasonal breathing of carbon between ocean and atmosphere—the sideways transport is the point. The spiral is one of the cleanest ways to see why that sideways transport appears at all.

There is also a deeper, almost philosophical implication: the ocean stores memory in its vertical structure. When wind blows, the surface responds quickly; the layers below respond more slowly, and they keep some trace of the forcing after the wind changes. In that sense, the spiral is not just a snapshot of competing influences but a mechanism by which short-lived atmospheric events can leave a longer-lived imprint in the upper ocean. A storm can deepen mixing, strengthen or weaken stratification, and reorganize the angle-by-angle distribution of momentum. Later, under calmer winds, the ocean's response is shaped by what the storm left behind. The sea surface may look calm, but the staircase remains, rearranged.

For microscopic life, this matters in a way that is easy to overlook. Phytoplankton live in the photic zone, the upper layer penetrated by enough sunlight to power photosynthesis. Too deep and they starve for light; too shallow and they may be battered by ultraviolet exposure or run out of nutrients. Many of them cannot swim in the way larger animals do; they drift and adjust buoyancy, but they are largely at the mercy of water motion. A wind event that thickens the mixed layer can entrain phytoplankton deeper, diluting them through a larger volume and reducing their average light exposure. A wind pattern that causes divergence can pull nutrient-rich water upward, offering a brief feast. The same physical mechanism that moves ice at an angle can, by subtle rearrangements of depth and mixing, change the odds of a bloom.

The staircase image also corrects a common misconception about the ocean as a two-dimensional surface with motion painted on top. Much of what the sea does happens in the vertical: the transfer of momentum downward, the compensating movements upward or downward required by convergence and divergence, the mixing that redistributes heat and dissolved gases. The Ekman spiral is a reminder that even wind-driven flow is not merely a surface phenomenon but a three-dimensional negotiation between forcing and rotation.

Stand again on that pier, watching the wind scud across the water. The surface may be streaked with foam, the waves marching in one direction. But beneath, the motion is not a single marching column. It is a turning, diminishing set of flows, each layer tugging on the next, each one obliged to turn because the whole planet is turning. Over hours and days, that layered response adds up to a sideways push that can empty a coastline's surface waters, pile water into a gyre, or quietly rearrange the thickness of the sunlit habitat. The sea does not simply go where it is pushed; it turns the push into structure.

1.2 The Great Vertical Pump

A wind can push the sea sideways for days and still leave the surface looking deceptively unchanged. The water does not pile up into a visible ridge; it usually slips away, spreading its departure over tens of kilometres. Yet the ocean cannot tolerate a sustained loss of surface water without paying it back. If water is carried off in one direction, something must replace it. The repayment is often vertical, and it arrives from below: darker, colder water rising into the sunlit layer, carrying with it a chemical inheritance from the deep.

Upwelling is the name given to that return flow. It sounds like a local event, something that happens in a bay or along a coastline, but it is better imagined as plumbing. Winds do not merely ruffle the surface; by driving Ekman transport they create broad regions where surface water converges (crowds together) or diverges (pulls apart). Where water diverges, the surface layer thins and the ocean draws up water from depth. Where it converges, the surface layer thickens and water is pressed downward. In either case, the vertical motion is rarely dramatic if measured as speed. A few metres per day is already lively. But the consequences are large because the ocean is stratified: properties change rapidly with depth, so a small vertical displacement can produce a large change in temperature, nutrients, oxygen, and even acidity.

The essential mechanism can be reached through a simple thought experiment at the scale of a town map. Picture a straight coastline. The wind blows parallel to it, steady enough that the upper ocean has time to settle into a wind-driven balance. In the Northern Hemisphere, the net transport in the surface boundary layer is to the right of the wind. If the wind blows along the coast with land on the left and open ocean on the right, the net transport is directed offshore. A strip of surface water is continually exported away from the coast.

Where does replacement water come from? It cannot come through the land, and it cannot come fast enough from along the coast if the wind is uniform. The only remaining direction is upward. Water rises along the continental shelf and slope, tilting into the light like a curtain being lifted.

This is coastal upwelling in its basic form: a mechanical linkage between alongshore winds and vertical motion. In the Southern Hemisphere the geometry flips, not because the wind changes, but because the Coriolis turning does. The wind may blow the same way along a similarly oriented coast, but the ocean's response is mirrored; the Ekman transport is to the left of the wind instead of the right. The pump changes direction simply because Earth's rotation changes sign across the equator.

Along the eastern edges of the world's ocean basins, this pump turns entire regions into biological powerhouses. The coasts of California, Peru and Chile, northwest Africa, and southwest Africa are famous for it: cold tongues of water, persistent coastal fogs, and fisheries that can be astonishingly productive. The pattern is not an accident of local geography. It is the marriage of winds, rotation, and the existence of a coastline that blocks the sideways escape route.

A concrete example can be felt in summertime along the U.S. West Coast. Many days bring northwesterly winds that run down the coast. The surface ocean, nudged by Ekman transport, is pushed offshore. Water rising from below is colder, often much colder than the air a few kilometres inland. The result is a familiar coastal contradiction: a hot interior and a chilly shoreline. The cold surface water encourages fog by cooling the air just above it, sometimes forming a bright white wall that rolls toward land. People experience the pump as weather. Seabirds and fish experience it as food.

The food arrives because deep water is not merely cold; it is chemically different. In the sunlit surface layer, phytoplankton consume nutrients such as nitrate, phosphate, and silicate as they grow. When they die, or when they are eaten and repackaged into sinking particles, some of that material falls out of the surface layer and is broken down at depth by microbes. The deep ocean becomes a storehouse of nutrients and dissolved carbon. Without a mechanism to return those nutrients upward, surface waters can become a kind of desert: brilliantly lit but nutritionally impoverished. Upwelling acts like an elevator, bringing the deep storehouse back within reach of photosynthesis.

This is why coastal upwelling zones are often called the ocean's breadbaskets. The phrase can mislead, because breadbaskets are usually stable. Upwelling ecosystems are not stable in that sense; they are dynamic, sometimes violently so. The wind pulses, the coastline funnels currents, and the food web responds with blooms, crashes, and sudden appearances of fish. The pump is a delivery system, but it delivers in bursts.

There is also a darker side to the delivery. Deep water can be low in oxygen because respiration at depth consumes it, and it can be rich in dissolved carbon dioxide because decomposition produces it. When such water rises onto the shelf, it can stress marine life. In some regions, the oxygen content can be low enough to create hypoxic conditions, and the elevated carbon dioxide can lower pH, contributing to local acidification. The same winds that feed fisheries can, under the wrong combination of conditions, bring stressors to the surface. The pump does not discriminate; it lifts whatever chemistry the deep currently holds.

The coastal picture is vivid because the boundary is visible. Yet the vertical pump is not confined to coastlines. In the open ocean, winds often vary across space. One patch of sea may be under stronger winds than another; the wind direction may rotate around a pressure system. When the wind stress changes from place to place, the Ekman transport changes as well, and that creates convergence or divergence without any coastline at all.

A useful way to think about this is to stop following the wind and start looking at its curl. Curl, in this context, is a measure of how much the wind stress field tends to make the surface flow rotate or shear across a region. If the wind stress curl is such that Ekman transport spreads outward, surface water diverges and water must rise from below. If the pattern squeezes surface water together, it converges and is pushed downward. The vertical velocity associated with this mass balance is called Ekman pumping (for downward motion in many conventions) or Ekman suction (for upward motion), though the names are used inconsistently enough that it is safer to keep the physical picture in mind: divergence pulls up; convergence pushes down.

At the scale of entire basins, these patterns organize the ocean's interior. The broad subtropical gyres in each ocean are not merely surface swirls; they are three-dimensional structures shaped by the fact that winds over the subtropics tend to drive convergence. Water is slowly pressed downward in the gyre interiors, thickening the warm surface layer and deepening the thermocline, the region where temperature drops rapidly with depth. The gyre centres become warm, nutrient-poor, and remarkably clear, the blue deserts of the sea. Farther poleward, in many regions, the winds favour divergence and upwelling, helping to maintain cooler surface waters and a shallower thermocline. This is a basin-scale pump: not a dramatic jet of water, but a persistent vertical rearrangement that sets the stage for climate and ecosystems.

The historical moment when these ideas were knitted into a global picture is often associated with Harald Sverdrup's work (1947). Sverdrup connected wind patterns to the interior transport of the ocean through a balance of vorticity, a measure of rotation in the flow. The details belong to geophysical fluid dynamics, but the emotional takeaway is simple: the wind does not just push water; it imposes a pattern of twisting and stretching that the ocean must answer with large-scale currents and vertical motions. Ekman's boundary layer provides the entry point for wind momentum. Sverdrup's insight helped show how that entry point communicates with the ocean interior, shaping the great gyres and their slow vertical breathing.

None of this is fast. The vertical velocities associated with basin-average Ekman pumping can be tiny compared with the speed of surface currents. They are often measured in metres per month rather than metres per second. That can sound irrelevant until one remembers that biology works on accumulation. A phytoplankton cell does not need a torrent of nutrients; it needs a reliable supply that offsets what is being consumed and exported. Likewise, climate does not require the ocean to overturn overnight; it requires the ocean to move heat and carbon on the timescales that matter to seasons, decades, and centuries. A pump that lifts only a small amount each day, but does so across millions of square kilometres, becomes a planetary machine.

Here is a way to feel the scale without losing the intimacy. Imagine a broad region of ocean where surface waters are diverging gently, as if the sea surface were a slow-motion split seam. Water rises from below at, say, a metre per day. That number is not universal, but it is within the right order of magnitude for energetic regions. A metre per day is imperceptible to a swimmer; it is slower than the settling of many marine snow particles and far slower than the motions driven by waves. But it is enough to lift a parcel of water from 100 m depth into the surface layer in a few months. In many parts of the ocean, the difference between those depths is the difference between nutrient-poor and nutrient-rich water, between low and higher carbon dioxide, between oxygenated and less oxygenated. The pump is slow, but it crosses steep gradients.

Coastal upwelling makes these gradients tangible. Stand on a cliff above an upwelling coast and look at the sea surface pattern. Cold upwelled water can appear as darker bands or filaments, sometimes traced by lines of foam or by the distribution of seabirds that have learned, in their own embodied way, the geometry of nutrients. Satellites reveal these patterns in sea-surface temperature: tendrils of cold water stretching offshore, spun into spirals and rings by instabilities. Those filaments are part of the pump's delivery network. They move the product of upwelling—nutrient-rich water and the blooms it fosters—away from the coast and into the wider ocean.

The pump also interacts with human economies in ways that are as immediate as dinner. The Peruvian anchoveta fishery, for example, has long been linked to the productivity of the Humboldt Current upwelling system. When winds and ocean conditions support strong upwelling, the food web can support huge fish biomass. When the system is disrupted, notably during strong El Niño events, warmer surface waters suppress nutrient supply to the surface and the ecosystem reorganizes. The mechanism is not only coastal Ekman transport; equatorial dynamics and changes in stratification matter too. But upwelling is a central thread: a wind-linked supply line that can be strengthened or weakened, with consequences that travel quickly from physics to plankton to fish to markets.

A historical anecdote that captures the scientific imagination behind upwelling is not a single dramatic expedition so much as a recurring surprise in measurements: the discovery that the ocean's most productive regions can be the coldest at the surface. Early oceanographers and fishermen alike learned that cold water near a coast is often a sign of life, not of sterility. It is the signature of deep water surfacing. In an era when warmth was often associated with fertility on land, the ocean offered an inversion: productivity could be driven by the rising of cold water, precisely because cold, deep water carries the nutrients that warm surface water lacks.

There is an additional subtlety that makes the pump feel algorithmic rather than purely mechanical. The ocean does not respond to wind forcing as a single slab; its response depends on the stratification and on the turbulence that mixes layers. Stratification is itself shaped by heat fluxes (sunlight, evaporation, cooling) and freshwater inputs (rain, rivers, ice melt). In a strongly stratified surface layer, the wind's momentum may not penetrate as deeply, and the vertical exchange can be constrained. In a weakly stratified or storm-mixed layer, the wind can stir deeper, altering the effective depth at which mass balance demands vertical motion. The pump's efficiency is therefore conditional: the same wind stress pattern can produce different vertical exchange depending on the ocean's internal state.

This conditionality becomes particularly vivid when considering smaller-scale turbulence. Mesoscale eddies and submesoscale fronts, swirling and slashing through the ocean like weather systems within the sea, can create vertical velocities that are locally much larger than the basin-average Ekman pumping. Patrice Klein and Guillaume Lapeyre (2009) argued for the importance of such turbulence-driven vertical exchange because tracers and nutrients change sharply with depth; intense but intermittent vertical motions can therefore dominate local budgets. James McWilliams (2016) emphasized how submesoscale currents, often associated with sharp density fronts, can be strongly time-dependent and capable of rapid vertical transport.

The existence of these turbulent routes does not replace the wind-driven pump; it complicates it. The wind patterns set the broad conditions: where the surface tends to converge or diverge, where the thermocline is deep or shallow, where the ocean is primed for nutrient scarcity or surplus. Eddies and fronts then modulate the delivery, sometimes enhancing it, sometimes trapping nutrients in rotating lenses, sometimes exporting coastal productivity offshore. The pump, in other words, has both a steady motor and a noisy gearbox.

Even with those complications, the core logic remains elegantly constrained by mass conservation. If surface waters are pushed away from a region for long enough, something must rise. If they are pushed into a region, something must sink. This is not a metaphor; it is accounting. The ocean cannot create water out of nothing and cannot compress it much. The vertical pump is the ocean's way of keeping its books balanced while responding to winds on a spinning planet.

For the smallest organisms, this accounting is destiny. A phytoplankton bloom is often described as if it were merely a biological event, a population explosion. But a bloom is also a temporary alignment of physics and chemistry: light is available, nutrients arrive, and the water remains in a state that allows cells to stay in the photic zone long enough to reproduce. Upwelling can deliver nutrients while also increasing turbulence, which can mix cells deeper. The outcome depends on timing and on the details of stratification. Some blooms erupt in the immediate wake of upwelling when winds relax and the surface layer restratifies, trapping nutrients and cells in a bright, stable layer. In that sense, the pump does not simply lift nutrients; it sets a stage on which light and mixing negotiate the terms of growth.

The sea surface above an upwelling region often carries a paradoxical sheen: cold water, sometimes capped by fog, beneath bright sunlight. It can feel austere from a beach, as if the ocean has turned its back on warmth. Yet just under that surface, the chemistry is being rebalanced, and with it the possibilities of life. The pump is not loud. It does not announce itself with a splash. It announces itself with a change in taste for the organisms that can sense nutrients, with a change in the colour of the water when chlorophyll increases, with the sudden presence of fish where there had been none, and with the thin, persistent insistence that the deep ocean is never entirely out of reach of the light.

1.3 Storms Within the Sea

A calm-looking ocean is often a lie told by distance. From a cliff or an airplane window the sea can appear like a single sheet, occasionally wrinkled by wind. Even from a ship, the eye is easily captured by waves and whitecaps, the quick, noisy motions at the surface. Yet much of the ocean's real restlessness is slower and larger, with shapes that do not announce themselves by sound. It takes instruments, patience, and a willingness to think of the sea as a kind of atmosphere turned upside down to notice them: rotating weather systems made of water, hundreds of kilometres across, lasting not hours but months.

These are mesoscale eddies. The word *mesoscale* is a scientist's compromise, meaning "in the middle": larger than the turbulent swirls you might see in a tidal rip, smaller than the great gyres that occupy whole basins. A typical mesoscale eddy has a radius on the order of 100 km and can persist for months. That scale is not a coincidence; it is a fingerprint of how rotation and stratification set limits on what kinds of motion can hold together. An eddy is not merely a whirlpool. It is a coherent parcel of ocean, often with its own temperature and salinity identity, that can carry heat, salt, nutrients, and living organisms across long distances while resisting being mixed away.

One way to make eddies emotionally real is to compare them to storms in the air. A thunderstorm is intense but fleeting and often localized; a mid-latitude cyclone can be a thousand kilometres wide and live for a week. In the ocean, the analogous systems are slower. Their winds are absent; their pressure fields are expressed not in millibars but in subtle changes of sea-surface height, often measured in centimetres. Yet they can have the same logic of rotating balance, the same ability to trap and transport, and the same tendency to organize the chaos around them into patterns.

The historical turning point came when the ocean acquired eyes in space. Before satellites, eddies were known from ship measurements and from drifting instruments, but they were hard to map. You could cross an eddy without realizing it, or you could infer it indirectly from temperature and salinity profiles. Satellite altimetry changed that by measuring sea-surface height with extraordinary precision. Because the ocean surface tilts slightly where currents flow-water piles up by a few centimetres on one side and dips on the other-sea-surface height becomes a proxy for pressure, and pressure gradients drive currents. With years of satellite sea-surface height stitched into global maps, eddies stopped being occasional curiosities and became an abundant, repeating texture of the sea. Dudley Chelton and Michael Schlax and colleagues (2011) used such altimeter records to track coherent eddies around the world, showing not only how many exist but how long-lived and mobile they can be. The ocean, viewed this way, looks less like a conveyor belt and more like a field of spinning cells.

A useful first question is why eddies form at all. Winds do important work at the surface, and Earth's rotation organizes much of the large-scale flow into geostrophic balance: a state where pressure-gradient forces are balanced by the Coriolis deflection, producing currents that tend to flow along lines of constant pressure rather than straight downhill. That balance can create remarkably steady-looking currents when averaged over long times. But the ocean is layered, and those layers can store energy in their tilts and slopes. When lighter water sits next to heavier water, or when the thermocline is sloped across a region, there is available potential energy waiting to be released. Under the right conditions, the layered ocean becomes unstable in a very particular way: it converts that stored potential energy into kinetic energy, spinning up eddies. Gill, Green, and Simmons (1974) articulated key aspects of how baroclinic instability-instability arising from density stratification-can generate eddies at the observed scales in a rotating ocean.

Baroclinic instability can be made intuitive without equations by thinking about a tilted stack of fluids. If a boundary between warm light water and cold dense water is slanted, gravity would prefer to flatten it. But rotation resists that flattening by constraining how water can move. The compromise is not a simple slump. Instead, the system sheds its tilt by peeling off rotating blobs and filaments-eddies-that carry warm water poleward and cold water equatorward, gradually smoothing the contrast. In the atmosphere, similar instabilities generate the familiar chain of low-pressure systems that march across weather maps. In the ocean, the same instability is slowed by water's density and by the ocean's strong stratification, producing eddies that are larger and longer-lived.

Eddies matter because they are the ocean's method of moving things that large-scale currents cannot easily move on their own. Heat is an obvious example. The ocean absorbs enormous amounts of heat from the atmosphere, and where that heat ends up depends on both mean currents and eddies. An eddy can carry a warm patch into cooler surroundings or vice versa, acting like a mobile thermos. Salt and freshwater also hitch rides, reshaping stratification and thereby influencing how deeply the wind can mix the surface. Nutrients, too, can be transported, but here the story becomes more delicate because nutrients are not just carried sideways; they are also delivered vertically.

A mesoscale eddy can lift or depress the thermocline. In a cyclonic eddy in the Northern Hemisphere (spinning counterclockwise), the sea surface is often slightly lower in the centre, and the pressure field is such that the thermocline domes upward. That doming can bring nutrient-rich waters closer to the sunlit layer, supporting phytoplankton growth. An anticyclonic eddy (clockwise in the Northern Hemisphere) often has a raised sea surface and a depressed thermocline, tending to be nutrient-poor near the surface. These tendencies are not rigid rules, because eddies interact with winds, with background currents, and with the coastline, but the basic idea is powerful: eddies are not merely horizontal swirls; they are three-dimensional distortions that can change the distance between deep nutrients and surface light.

That three-dimensionality is where the earlier picture of a wind-driven vertical pump meets turbulence. A broad upwelling region can deliver nutrients to the surface, but mesoscale eddies and fronts decide how that delivery is packaged and where it goes. The upwelled water does not spread like ink in a still glass. It is stretched into filaments, rolled into spirals, and pinched off into rotating lenses. An upwelling event can therefore generate a kind of biological architecture: ribbons of high chlorophyll that can be seen from space, moving and reshaping over days and weeks. For a phytoplankton cell, the ocean is not a uniform soup but a shifting mosaic of feast and famine.

A concrete example can be found in the eddy field of the Southern Ocean, the ring of water encircling Antarctica. There, strong westerly winds and steep density gradients create one of the most energetic eddy environments on Earth. Mesoscale eddies are so common that they become a defining feature of the region's circulation. They also play a large role in how heat and carbon are exchanged between the surface and the interior. Warm waters can be carried southward and upward by eddies in ways that complicate simple pictures of a steady current. For the organisms living there, from phytoplankton to krill, eddies can create transient oases where nutrients and light briefly align.

Another vivid setting is the Gulf Stream region off the eastern United States. The Gulf Stream is often imagined as a narrow river in the sea, but in reality it is accompanied by a lively entourage of meanders and rings. Large warm-core rings can pinch off from the current and drift into colder waters; cold-core rings can form on the other side, injecting cooler, nutrient-rich water into warmer surroundings. These rings can persist for months, carrying their distinct water properties and biological communities with them. Fishers have long been aware that such features can concentrate life, influencing where fish aggregate, where birds feed, and where catches can be made. The language of “edges” and “fronts” used by people who work at sea often refers to these dynamic boundaries shaped by eddies.

The fascination of eddies is also that they force humility. One might hope to write a clean energy budget: wind puts energy into the ocean; mean currents carry it; turbulence dissipates it. But the ocean does not separate its processes neatly. Ferrari and Wunsch (2009) emphasized how difficult it can be to distinguish between motions that are directly forced, motions that are part of a turbulent cascade, and motions that are wave-like. Internal waves, tides interacting with rough topography, and instabilities of currents all mingle. Mesoscale eddies sit in the middle of this confusion, sometimes behaving like turbulence, sometimes like organized structures, sometimes interacting strongly with waves. The ocean’s “storms” are not just messy; they are structurally ambiguous.

This ambiguity becomes even sharper when the scale shrinks below mesoscale to what is called *submesoscale*. These are fronts, filaments, and small vortices with horizontal scales of roughly 1 to 10 km, often living for days rather than months. Their existence has been known for a long time to those who have watched slicks and foam lines on the sea surface, but only relatively recently have they been observed and modeled systematically enough to be recognized as a central part of upper-ocean dynamics. James McWilliams (2016) described submesoscale currents as a frontier because they are often ageostrophic: they do not obey the neat geostrophic balance that organizes much of the mesoscale. They can generate strong vertical velocities and intense mixing, providing a pathway for energy to cascade toward dissipation and for surface properties to be injected downward.

Why does this matter for the idea of a vertical pump? Because the vertical exchange that matters to biogeochemistry is often not the basin-average upwelling rate but the actual pathways by which parcels of water cross density layers. Mesoscale and submesoscale motions can create localized, intense vertical transport events—brief pulses that can dominate the experience of plankton and microbes. Klein and Lapeyre (2009) argued that these turbulent processes can induce a vertical pump of their own, one that is intermittent and spatially patchy, and therefore difficult to observe and to average meaningfully. A satellite can reveal a chlorophyll spiral, but it cannot directly measure the vertical velocities that fed it. A float can measure temperature and salinity, but it drifts and samples a narrow path. The ocean keeps its best secrets by distributing them across scales.

Still, some robust qualitative truths can be carried to a dinner table without apology. Mesoscale eddies are not rare exceptions; they are ubiquitous. They are not superficial; they have depth and structure. They are not merely noise; they transport heat, salt, carbon, and life in ways that shape the ocean's role in climate. They can trap a patch of water-along with its planktonic community-and carry it across environmental gradients, exposing organisms to new light regimes, temperatures, and predators. They can also act as boundaries that keep communities apart, maintaining distinct ecological provinces the way mountain ranges and river systems can structure life on land.

One of the most striking ecological consequences is the creation of mobile habitats. A coherent eddy can be a moving island. In its core, conditions can be relatively stable compared with the surrounding waters, at least for a time. Nutrients may be slightly higher, or the mixed layer slightly deeper, or the temperature slightly warmer. Zooplankton and fish larvae can be retained, recirculating rather than being swept away. If the eddy interacts with another feature or with topography, it can shed filaments-thin streams of water that export organisms and nutrients like spores from a plant. The ocean's biodiversity, often discussed as if it were spread smoothly across latitudes, is in fact constantly being rearranged by these mobile islands and their fraying edges.

There is also an aesthetic to eddies that is easy to miss if one only thinks in graphs. In satellite images of sea-surface temperature or chlorophyll, eddies appear as spirals and commas, as if the ocean is constantly sketching calligraphy. Those shapes are not decoration; they are the visible trace of how a rotating fluid stretches and folds. A filament of productive coastal water can be pulled offshore and wrapped around an eddy, creating a spiral band of green in an otherwise blue region. Over time the filament thins, and mixing blurs it, but for days or weeks it can remain coherent enough that one could almost point to it on a map and say: there, the water is different. There, the food web has been given an address.

It is tempting to end with a promise that ever-better satellites and models will tame this complexity into a predictable forecast, an ocean weather report as routine as the atmospheric one. Progress is real, but the ocean's turbulence is not an enemy to be conquered; it is a form of organization that emerges precisely because the ocean is rotating, stratified, and forced at its boundaries. Even perfect observation would not eliminate the inherent sensitivity of instabilities and the chaotic interactions among eddies, fronts, and waves. What it can do is change the questions from “Is the ocean steady?” to “What patterns of variability are typical, and what do they make possible?”

Along the edge of an eddy, the sea can look ordinary. A few kilometres away, the temperature may differ by a degree or two; the chlorophyll may be doubled; the mixed layer may be tens of metres deeper. For a human in a boat, this is a mild gradient. For a microbe or a phytoplankton cell, it can be the difference between running out of nitrogen and finding a surplus, between being mixed into darkness and lingering in light. The ocean’s storms do not always break the surface, but they redraw the living world all the same.

Chapter 2

Sunlight and the Invisible Forest

The ocean surface hides a botanical engine that rivals the Amazon in carbon capture, yet possesses no trunks, leaves, or roots. We explore how microscopic algae navigate a perilous vertical gauntlet, balancing the metabolic need for solar energy against the relentless drag of gravity and the chaotic mixing of the mixed layer.

2.1 The Green Machine: Photosynthesis in Fluid

A leaf can afford a kind of certainty. It grows from a branch, spreads into sunlight, and—unless something tears it away—stays put. A phytoplankton cell gets no such bargain. It lives in a restless medium where even "up" and "down" are negotiated every second by turbulence, sinking, and buoyancy. Yet these drifting cells perform one of the planet's most consequential tricks: they turn light into living matter, and in doing so help set the tempo of fisheries, the chemistry of the air, and the ocean's long-term storage of carbon.

The word *phytoplankton* sounds like a single thing, but it is more like a crowded city: diatoms in glassy shells, nimble flagellates with whips for propulsion, tiny cyanobacteria that behave like bacteria and photosynthesize like plants, and stranger forms that blur categories. What unites them is not their shape but their job. They are primary producers-organisms that build organic carbon from carbon dioxide using sunlight. Everything else in the ocean food web ultimately leans on that conversion.

Photosynthesis is often described as if it were simply "powered by light," as though a cell were a solar panel. In practice it is a careful accounting problem: photons arrive in uneven bursts, chemistry has a pace, and life has running costs. Every phytoplankton cell continuously pays for maintenance-repairing proteins, pumping ions across membranes, keeping its internal chemistry within tolerable bounds. Those costs show up as *respiration*, the biochemical process that consumes some of the carbon and energy previously fixed by photosynthesis. If photosynthesis outpaces respiration, the cell can grow and divide. If not, it shrinks, waits, or dies.

That balance is easiest to picture with two depths. The first is the *compensation depth.* Imagine a single cell held at one depth long enough to settle into a routine: its photosynthesis depends on how much light reaches that level, while its respiration is relatively steady. At shallow depths, light is plentiful and photosynthesis exceeds respiration. Deeper down, light thins out until there is a point where the two are equal: no net gain, no net loss. Below that compensation depth, a cell living there steadily loses ground unless it has stored energy or a special strategy.

The second depth belongs not to one cell but to a population being shuffled up and down by mixing: the *critical depth.* This idea is tied to a historical moment when oceanography was learning to speak in the language of budgets and integrals rather than impressions.

In 1953, Harald Sverdrup—one of the architects of modern physical oceanography—published a short paper with an audacious goal: explain the sudden springtime surge of phytoplankton in high-latitude seas using a simple physical criterion. The dramatic spring bloom had long been known to sailors and naturalists; it could tint water green, clog nets, and feed a seasonal cascade of zooplankton and fish larvae. Sverdrup's insight was to treat the upper ocean as a mixed layer stirred so vigorously by wind and convection that phytoplankton were, in effect, passengers on an elevator—spending part of their day near the surface in bright light and part of it deeper down in dimness.

Sverdrup asked a deceptively plain question. If a population is thoroughly mixed from the surface down to some depth, can it, *on average*, grow? The answer depends on the depth of mixing. Over the upper portion of the water column, photosynthesis is strong; deeper, it weakens. Respiration, in Sverdrup's simplest framing, is a roughly depth-independent loss rate that applies no matter where the cell is in that mixed layer. So one can draw a curve of gross photosynthetic production decreasing with depth, and a horizontal line of losses. Where those two meet is the compensation depth for a single cell held still. But mixing does not hold cells still. Mixing spreads the population across bright and dark. The relevant question becomes: does the *integrated* photosynthesis across the mixed layer exceed the *integrated* losses across that same layer?

The critical depth is the mixed-layer depth at which those two integrals balance. If the ocean is mixed deeper than the critical depth, the population spends too much time in low light; total gains cannot pay total costs, and biomass declines. If the ocean mixes shallower than the critical depth, the gains from the well-lit part of the layer outweigh the losses accrued everywhere, and the population can increase.

This is the core of Sverdrup's critical depth hypothesis: as winter ends, surface waters often warm and freshen, stratification strengthens, and the mixed layer becomes shallower. When the mixed layer shoals above the critical depth, a bloom becomes possible. It is a stark, elegant picture: physics sets the habitat's vertical extent, and biology responds.

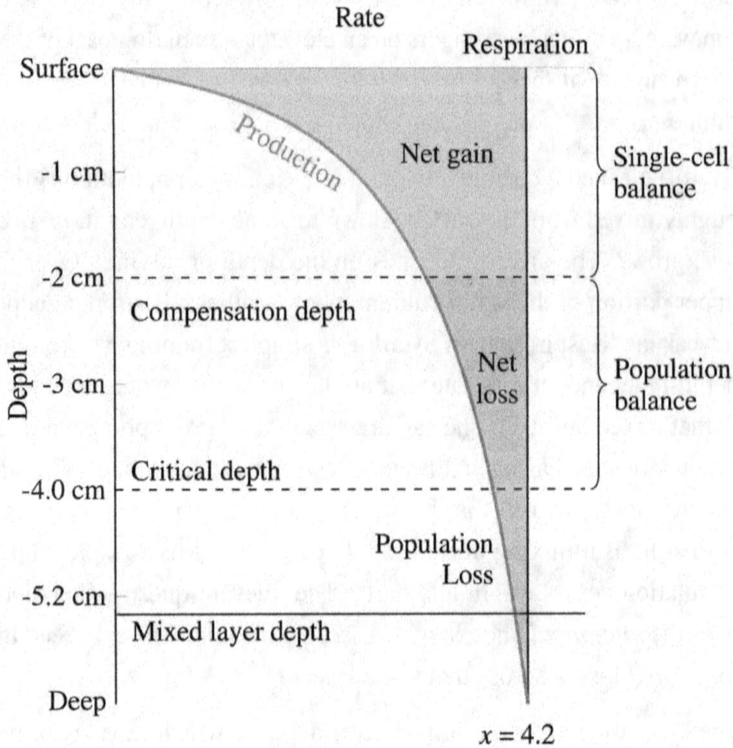

A diagram like this can mislead if it feels too tidy. In the sea, neither "production" nor "respiration" is a single curve, and "mixed layer" is not a rigid box with a lid. Still, Sverdrup's picture remains powerful because it turns a vague seasonal story into a quantitative one: the upper ocean is a light field with costs, and mixing determines how organisms sample that field.

To understand why the balance is so touchy, it helps to imagine what a phytoplankton cell is actually experiencing. Light underwater is not merely dimmer with depth; it is transformed. Water absorbs red wavelengths strongly, so the spectrum becomes bluer as you go down. Particles scatter light, making the direction of photons less orderly and the environment more diffuse. The rate at which light decays with depth is often approximated as an exponential decline, which means the first few meters can matter enormously. A cell at two meters might be living in a world twice as bright as a cell at ten, and ten might be twice as bright as thirty. If mixing sweeps a cell through that range, it is not just changing the mean light; it is imposing a flicker—bright, dim, bright—on timescales that can be seconds to hours.

Cells do not respond to light like instant-read thermometers. The machinery that captures photons and turns them into chemical energy has inertia. Pigment-protein complexes in the membranes must be assembled and repaired; enzymes in the carbon-fixing pathway must be activated; excess energy must be safely dissipated to avoid damage. A useful everyday analogy is not a solar panel but a kitchen. A sudden influx of guests does not immediately produce a smooth stream of meals. The cook scrambles, the stovetop bottlenecks, and at some point more ingredients do not help because the limiting factor is the speed of preparation. Likewise, photosynthesis often rises with light at low intensities, then saturates when the cell's biochemical throughput is maxed out. At very high light, some organisms suffer *photoinhibition*—damage or down-regulation when photons arrive faster than the cell can safely process.

Now place that kitchen on an elevator. A turbulent eddy can lift a cell toward the surface, giving it a burst of bright light that might saturate photosynthesis, then drop it back into dimness where photosynthesis slows but respiration keeps ticking. Some organisms cope by adjusting their pigment content—packing more chlorophyll into their light-harvesting systems when they expect dim conditions, and reducing it when they live near the surface. But pigment is not free. Chlorophyll and accessory pigments require nutrients, especially nitrogen, and building them takes time. There is an economy here: a cell that invests heavily in light capture can do well in deep or cloudy waters, but may become vulnerable near the surface where high light can turn that large antenna into a liability.

The physics adds a second kind of constraint: not just light variability, but the simple fact that cells sink. Many phytoplankton are slightly denser than seawater. A diatom's ornate silica shell is beautiful under a microscope, but it is also ballast. Some diatoms can regulate their density a bit using ions or stored compounds; some can change shape or make spines that increase drag. Others rely on turbulence to keep them suspended. The problem is size-dependent. Very small cells experience the ocean as a thick fluid; viscous forces dominate and sinking is slow. Larger cells and colonies sink faster, and their need for mixing becomes more acute.

Even for a cell that could, in principle, photosynthesize rapidly near the surface, sinking means it may spend too much time below the compensation depth unless turbulence continually rescues it. This is where Sverdrup's emphasis on a "thoroughly mixed" layer matters. If mixing is intense, cells are redistributed quickly and no one stays deep for long; but intense mixing often goes hand-in-hand with deep mixing, which can lower the mean light exposure. If mixing is weak, a shallow layer might have abundant light on average, but individual cells might sink out of it and be lost from the bright habitat. The ocean sets a moving target: a favorable mean light can coexist with unfavorable transport, and vice versa.

A third constraint sits at the cell's surface, where physics and physiology literally touch. In seawater, nutrients such as nitrate, phosphate, and iron arrive at a cell by diffusion and by the stirring of the surrounding water. Around every cell there is a thin *boundary layer*—a region where water motion slows and diffusion does more of the work. For a tiny cell, diffusion can be enough; for a larger one, or for a fast-growing one, the delivery rate can become limiting. Turbulence can help by thinning boundary layers and refreshing the water around cells, increasing nutrient supply. Yet turbulence can also disrupt delicate chains of diatoms or alter encounter rates with grazers. The same stirring that helps feed a cell can also help its predators find it.

These competing effects make "more mixing" neither simply good nor simply bad. They also explain why there are multiple ways to refine Sverdrup's original criterion. Some later work distinguishes the depth of the hydrographic mixed layer (where temperature and density look uniform) from the depth of active turbulence (where water is actually being stirred hard enough to move cells). In some conditions, the ocean can appear mixed in density while turbulence is weak; in others, turbulence can penetrate below the obvious density boundary. For a phytoplankton population trying to balance photosynthesis against losses on timescales of a day or so, that distinction can matter as much as a change of tens of meters in mixed-layer depth.

It is also worth remembering that "losses" are not only respiration. Phytoplankton are eaten—often spectacularly efficiently. A spring bloom is not merely a plant party; it is a race against the mouths that wake up and multiply when food becomes abundant. Grazing pressure can be low in early spring if zooplankton populations are still small or cold-slowed, which makes it easier for phytoplankton biomass to accumulate once light conditions improve. As the season advances, grazers can catch up, and the same phytoplankton growth that once produced a visible bloom can be tightly cropped, leaving less chlorophyll even if photosynthesis is still high. In that sense, Sverdrup's elegant light-versus-respiration balance is a necessary part of the story but not always a sufficient one. The ocean's invisible forest does not simply "turn on" when light permits; it is constantly negotiated among physiology, physics, and predation.

A concrete example of how sensitive these balances can be comes from high-latitude waters at the end of winter. Imagine the North Atlantic in late February: storms repeatedly punch the ocean surface, deepening the mixed layer and sending phytoplankton on long excursions into dim water. Day length is increasing, but sunlight is still weak and often filtered through clouds. In Sverdrup's framing, the deep mixed layer keeps the population below the critical depth, so biomass should decline. Yet satellites and field measurements have sometimes found signs of growth beginning while mixing is still deep. One way to reconcile this is to notice that storms do not only deepen the mixed layer; they also come and go. Between storms, turbulence can relax and the effective mixing experienced by cells can become much shallower than the density-defined mixed layer suggests. Another possibility is ecological: if grazers are strongly diluted by deep mixing or slowed by cold, the "loss" term drops, allowing net accumulation even when light is marginal. The lesson is not that Sverdrup was wrong, but that the ocean is allowed to be intermittent. Biology lives in the gaps between averages.

That intermittency is part of what makes phytoplankton feel less like "plants in water" and more like exquisitely tuned sensors. Many species carry photoprotective pigments and can dump excess energy as heat within seconds, a way of surviving brief trips to the surface on bright days. Some can swim—slowly, but enough to bias their position upward over hours if turbulence is weak, or to hover at a favored depth where light and nutrients balance. Others form resting stages, sinking to the seafloor when conditions are poor and reentering the surface ocean when mixing returns them to light. Diatoms can bloom rapidly when nutrients are plentiful and light becomes adequate, building silica shells that both protect and weigh them down; later they may sink en masse, exporting carbon to depth in a pulse that feeds deep-sea ecosystems and contributes to the long-term sequestration of carbon.

None of these tricks abolishes the underlying arithmetic. A phytoplankton cell is a carbon accountant operating in a flickering light field, in a fluid that is always rearranging the terms of its budget. What makes the ocean's primary production so formidable is not that conditions are stable, but that so many different organisms have evolved ways to make a living in different combinations of light, mixing, and loss. The sea does not offer a single "right" depth or a single "right" strategy. It offers a moving, layered gradient—and a ceaseless, planet-wide competition to turn passing sunlight into the substance of life.

2.2 Layers of Life: Stratification and Survival

Stand on a pier on a calm summer day and the sea can look like a single, continuous substance: the same water from surface to seabed, differing only in darkness and distance. From the perspective of a phytoplankton cell, that impression is almost comically wrong. The upper ocean is often arranged like a layer cake whose slices are defined not by taste but by density. A few degrees of temperature, a small change in salinity, even a thin lens of freshwater from a river or melting ice can create a barrier strong enough to reshape an ecosystem. Light enters from above and fades downward; nutrients tend to be replenished from below and become scarce upward. Stratification—the formation of stable density layers—pins these opposing gradients in place, creating a set of niches stacked vertically within the first few tens to hundreds of meters.

Density in seawater behaves in a way that surprises people who grew up thinking of "water" as a single thing. Warm water is lighter than cold water. Fresh water is lighter than salty water. Put warmth and freshness together at the surface and you get a buoyant cap that resists being mixed downward. Put cold, salty water below and you get a heavy base that resists being mixed upward. Wind can stir the surface and deepen the mixed layer, but a sufficiently strong density contrast acts like a lid: turbulence loses its grip, and the surface layer becomes a distinct habitat rather than a mere portion of a continuous column.

That lid is not just a physical boundary; it is an ecological one. Above it, phytoplankton live in a world of photons and relative nutrient poverty. Below it, nutrients are more abundant, but light is weak. The drama of many ocean regions is that primary producers need both, and stratification often separates them.

A useful pair of terms helps make this concrete. The *mixed layer* is the near-surface layer where temperature and density are relatively uniform because wind and surface cooling have stirred it. The *thermocline* is the zone where temperature changes rapidly with depth; in many places it coincides with a sharp density gradient. Biologists often talk as well about the *nutricline*, the depth range where nutrient concentrations increase strongly with depth. In the simplest picture, the base of the mixed layer sits above the nutricline: plenty of light up top, plenty of nutrients down below, and a frustrating gap between supply and demand.

The previous idea of a critical depth framed bloom potential as an average over a well-mixed layer. Stratification complicates that story by challenging the assumption that mixing is uniform, constant, and sufficiently vigorous to keep the population evenly distributed. A water column can be hydrographically "mixed" in the sense that temperature looks uniform over a certain depth range, while the actual turbulence that moves cells can be intermittent, patchy, and shallow. Conversely, a sharp density gradient can exist while occasional storms punch turbulence through it for hours at a time. For an organism that lives on the timescale of hours to days, those details are not footnotes; they are the habitat.

A historical anecdote captures how long people have noticed the biological consequences of layering, even before the physics was formalized. In the late nineteenth and early twentieth centuries, naturalists towing plankton nets began to report that "plankton" was not evenly distributed with depth; it formed layers, patches, and thin sheets that seemed to appear and vanish. Some of this was surely sampling bias—nets mix water as they move—but some was real structure. The ocean was telling observers, in the only way it could, that its biology had architecture.

Modern instruments made the architecture undeniable. Lower a conductivity-temperature-depth probe (a CTD) and you see density steps and gradients. Add a fluorometer, which measures the faint red glow emitted by chlorophyll when it absorbs light, and a striking pattern often emerges: a peak in chlorophyll not at the surface, but below it. This is the *Deep Chlorophyll Maximum* (DCM), a phrase that sounds like a technical detail until you realize it describes a hidden forest—an accumulation of photosynthetic pigment in a zone that is literally invisible from above.

The DCM tends to sit near the base of the mixed layer or just below it, often close to the top of the nutricline. That placement is no accident. Above the DCM, light is plentiful but nutrients are scarce. Below it, nutrients increase, but light fades quickly. The "sweet spot" is where cells can still photosynthesize, albeit slowly, and can also access enough nutrients to build the proteins and pigments that make photosynthesis possible.

To say there is a deep maximum of chlorophyll is not yet to say there is a deep maximum of life. Chlorophyll is both a pigment and a signal, and the signal can be tricky. When light is scarce, many phytoplankton respond by increasing the amount of chlorophyll per cell—like turning up the sensitivity on a camera in a dim room. A layer can therefore show a strong fluorescence peak because cells are *darker green*, not because there are more of them. Oceanographers sometimes distinguish a deep chlorophyll maximum driven primarily by photoacclimation from a deep *biomass* maximum in which there are genuinely more cells or more carbon. The difference matters for carbon cycling and food webs, but it also matters for something as everyday as a satellite map of "chlorophyll": satellites see only the near-surface optical layer and can miss much of the subsurface story.

In stratified waters, the surface can look deceptively barren even when the subsurface is busy. The Mediterranean Sea in summer offers a concrete example. Surface waters become strongly stratified and nutrient-depleted; the water can appear intensely blue, suggesting low chlorophyll. Yet a DCM often forms tens of meters down. It is a living compromise: enough light to keep the photosynthetic engine turning, enough nutrients creeping upward from depth (or delivered laterally) to make pigment and proteins, and enough stability that cells are not constantly mixed into darkness. A diver might never notice it, because the layer does not always change the water's appearance at the surface. A plankton net towed only near the surface might also miss it. But for grazers that can migrate vertically, or for carbon export processes that depend on where biomass accumulates, that hidden layer can be decisive.

Stratification does more than create a favorable depth for a chlorophyll peak; it shapes who thrives there. Consider cell size. Small phytoplankton have a high surface-area-to-volume ratio, which helps them absorb scarce nutrients. They also tend to sink slowly. In warm, stratified, nutrient-poor surface layers, small cells often dominate because they can survive on thin rations. Larger phytoplankton like many diatoms can grow rapidly when nutrients are abundant, but they sink faster and are more dependent on turbulence to keep them in the light. Stratification thus acts like a filter that changes the community, not just the total amount of chlorophyll.

There is also a mechanical reason stratification can matter even when light and nutrients are held fixed in imagination. Turbulence is not merely a blender; it has a structure that depends on density. When surface waters are strongly stratified, turbulence generated at the surface tends to die out more quickly with depth because buoyancy resists vertical motion. This can trap phytoplankton in the illuminated layer for longer periods, increasing their average light exposure even if the mixed layer, defined by temperature, remains similar. It can also trap grazers, larvae, and particles in the same zone, intensifying interactions—feeding, competition, infection—because everything shares the same thin habitat.

That thinness can be startling. Oceanographers sometimes observe *thin layers* of plankton only a few meters thick, extending for kilometers. These layers can form where stratification is strong and shear is present—where currents at different depths slide past one another. The physical details can be complex, but the ecological consequence is simple: concentrate prey into a narrow band and predators benefit. A layer can become a buffet line. It can also become a hotspot for reproduction, because encounters between cells and gametes, or between predators and prey, become more likely.

Stratification is often seasonal, and that seasonality carries the emotional rhythm of coastal life. In many temperate regions, winter storms and surface cooling deepen mixing, delivering nutrients upward but diluting phytoplankton through a thick, dark layer. Spring brings calmer weather and surface warming; stratification strengthens and the mixed layer shoals. Nutrients that were once stirred upward are now trapped below the surface cap, but the surface layer is brighter and more stable. In classic narratives of the spring bloom, this is the moment when phytoplankton can accumulate faster than they are lost—an ecological release. But even where that narrative holds broadly, stratification adds nuance: the bloom does not simply happen "at the surface." It happens in a particular vertical structure, with peaks that can sit near the surface early on and then migrate downward as nutrients are consumed above and the nutricline sharpens.

The idea of "mixed layer depth" can make this feel like a simple geometric problem—shallower equals better for light. Yet one of the most interesting debates in biological oceanography in recent decades has been about whether the depth of the hydrographic mixed layer is the right physical variable at all. Turbulence can be intense during storms and weak between them. The actively turbulent layer can be shallower than the mixed layer inferred from density. This opens the door to a different mechanism for bloom initiation: not the shoaling of the mixed layer, but the weakening of mixing. If turbulence drops below a certain level, cells can spend more time near the surface even if the water column still looks "mixed" by the usual hydrographic criteria. In other words, stratification is not the only way to create a shallow effective habitat; a calm spell can do it too.

This distinction is not academic nitpicking. It changes how one reads the ocean's calendar and how one anticipates ecological responses to climate change. Warming tends to strengthen stratification in many regions by heating the surface and increasing buoyancy, and increased freshwater input from rainfall, rivers, or melting ice can do the same. Stronger stratification can make it harder for deep nutrients to reach the surface, potentially reducing surface productivity in nutrient-limited subtropical gyres. At the same time, stronger stratification can stabilize the upper ocean, favoring small phytoplankton and shifting food webs toward smaller grazers. Meanwhile, in high latitudes where light is often limiting, a shallower, more stable surface layer can increase the time phytoplankton spend in light, potentially enhancing blooms—unless nutrients become limiting sooner or grazing dynamics shift. The same physical tendency can have different biological outcomes depending on what is limiting growth in the first place.

A concrete contemporary example comes from the Arctic and subarctic seas, where freshwater from ice melt can form a strong surface lens. That lens can create a very shallow mixed layer with ample light, which sounds favorable for phytoplankton. But the same lens can also isolate the surface from deeper nutrient reservoirs. Under sea ice, the story becomes even more delicate: light is filtered and reduced, but stratification can be intense, and the timing of ice retreat can shift the balance of light, mixing, and nutrient supply. In some years and places, this combination supports intense, short-lived blooms; in others it produces a quieter, prolonged subsurface maximum. For coastal communities and fisheries that depend on predictable food-web timing, these changes are not abstractions.

The DCM is perhaps the most vivid emblem of stratification's ecological creativity because it represents a solution to a contradiction: light above, nutrients below. But it is not a passive feature. It can actively influence the physics and chemistry around it. A concentrated layer of phytoplankton absorbs light, warming the water slightly and potentially reinforcing stratification on small scales. It alters oxygen and carbon dioxide locally through photosynthesis and respiration. It provides food that can attract migrating grazers, which in turn can recycle nutrients through excretion within or above the layer, tightening a local nutrient loop. In some cases, this recycling can help maintain the layer, like a small city sustaining itself with internal commerce.

There is a temptation to treat stratification as a stable scaffold and biology as paint applied to it. In reality, the scaffold flexes. Internal waves—gravity waves that travel along density interfaces—can lift and lower the nutricline and the DCM over hours. A layer that sits at 60 meters at dawn might be at 40 meters by afternoon, then sink again by night, not because the organisms are swimming en masse but because the density surfaces they are associated with are moving. If one were to sample at a fixed depth without knowing this, the biological signal would look like a daily boom-and-bust cycle. With the physical context, it becomes a moving habitat, sliding up and down like a slow elevator.

For phytoplankton, these vertical motions are not merely displacement; they are shifts in the terms of survival. A lift into brighter light can increase photosynthesis but also increase light stress. A drop into deeper, nutrient-rich water can provide resources for rebuilding pigments and proteins, but at the cost of lower photosynthetic income. Species differ in how they handle the trade. Some are sprinters that seize brief opportunities. Others are savers that store nutrients internally when they can, then draw on reserves when the environment turns lean. Stratification, by creating persistent gradients rather than uniform conditions, rewards these strategies and creates room for diversity.

One can see the effect of this diversity in the way different ocean regions “color” their subsurface layers. In oligotrophic gyres—vast regions of the subtropics that are famously clear and blue—the surface often contains low chlorophyll, but a DCM can still form where nutrients begin to rise. The dominant phytoplankton there are often tiny, adapted to scarcity, and their deep maxima may be driven strongly by photoacclimation. In contrast, in upwelling regions where nutrients are delivered to the surface by rising deep water, the surface can be rich and green, and the vertical structure can be more about light limitation and mixing than nutrient shortage. Coastal shelves, influenced by tides and variable freshwater, can flip between stratified and mixed states over days, producing biological responses that are correspondingly dynamic.

Stratification also shapes risk. A phytoplankton cell living near the surface in a stable layer is exposed not just to light but to ultraviolet radiation and to rapid temperature changes. Some species produce protective compounds; others rely on repair mechanisms. Viruses and bacteria that infect phytoplankton can also concentrate where hosts are concentrated, turning a productive layer into a battleground of infection. In a thin, stable habitat, a pathogen does not have to search far. The ocean's layering can therefore amplify not only growth but also collapse, producing sudden declines that look mysterious if one expects biology to respond smoothly to average conditions.

Seen this way, stratification is not simply "stability" versus "mixing." It is the ocean's way of organizing space into overlapping resource gradients, with interfaces that are both barriers and meeting places. The base of the mixed layer is an interface where sinking particles slow, where nutrient fluxes are focused, where internal waves ripple, where grazers can commute between light-rich feeding grounds and darker refuge. It is also where an invisible forest often thickens, not because it is the best of all possible worlds, but because it is the most tolerable compromise.

On a calm day, the surface can give the illusion that nothing is happening beneath. The real action may be a few tens of meters down, in a band so thin that a ship could pass over it without noticing, yet so biologically dense that it drives the daily meals of grazers and the slow export of carbon to the deep. A density gradient, drawn by differences in heat and salt, becomes an ecological boundary. In the ocean, physics does not merely set the stage; it partitions the audience, dims and brightens the lights, and decides where the food is delivered.

2.3 The View from Orbit: Planetary Chromatics

A sailor can tell rich water from poor water by eye. Some seas look like cobalt glass; others carry a greenish cast, or a milky turquoise, or a tea-stained brown from rivers. For centuries, those colors were treated as local character, like the smell of a harbor. The modern twist is that color became a measuring device: a way to estimate, from hundreds of kilometers above, how much microscopic plant life is floating near the surface.

That leap—from "the water looks green" to "there are about this many milligrams of chlorophyll per cubic meter"—is not a matter of clever photography. It is radiometry, the science of measuring light quantitatively. Satellites do not simply take pictures of the sea; they count photons (or, more precisely, measure radiance) in carefully chosen wavelength bands and try to infer what must be in the water to have produced that pattern.

The difficulty is that the ocean is a reluctant billboard. The signal from phytoplankton is subtle, and most of the light the satellite sees did not come from the ocean at all. Sunlight enters the atmosphere, scatters off air molecules and aerosols, reflects off clouds, and glints off waves. Only a fraction penetrates the surface, interacts with the upper ocean, and then escapes back upward. Of that escaping portion, only part reaches the satellite without being scrambled by the atmosphere. Ocean-color remote sensing is the art of extracting a small biological whisper from a loud atmospheric conversation.

The reason it is possible at all comes down to a fortunate mismatch between pure seawater and chlorophyll. Pure seawater absorbs red light strongly and lets blue light travel farther, which is why clear ocean water looks blue. Chlorophyll *a*—the central photosynthetic pigment shared across most phytoplankton—absorbs strongly in the blue and red parts of the spectrum and reflects or scatters more green. When phytoplankton are abundant, they steal some of the blue light that would otherwise return to space and allow more green to dominate. The ocean's color shifts, not dramatically like paint, but measurably.

A satellite sensor therefore watches the ocean through a set of spectral "keyholes," typically in the blue and green wavelengths, and sometimes additional bands that help with atmospheric correction or with distinguishing other substances. The sensor records the radiance leaving the top of the atmosphere in each band. What scientists want, however, is the *water-leaving radiance*: the light that actually emerged from the sea surface due to scattering within the water, after removing the atmosphere's contribution. That subtraction is the first major algorithmic step, and it is where much of the uncertainty begins.

It helps to picture the atmosphere as a veil that is sometimes thin and sometimes thick. In very clear air, Rayleigh scattering from molecules produces a predictable bluish haze. In dusty or smoky air, aerosols add extra scattering with a different spectral fingerprint. A satellite algorithm must estimate how much of the measured signal is due to these atmospheric effects and remove it. Over the open ocean, the challenge is especially severe because the ocean's own contribution in the visible bands is small compared to the atmospheric path radiance. Getting the subtraction slightly wrong can produce a plausible-looking ocean map that is systematically biased.

Despite these difficulties, the approach works well enough that "chlorophyll maps" have become a standard way of visualizing the living ocean. Those maps are not direct measurements of chlorophyll in the way a laboratory extraction would be. They are an inference produced by a chain of models: radiative transfer through the atmosphere, reflectance at the surface (including the messy physics of waves and foam), light scattering and absorption by water and particles, and finally a statistical relationship between spectral ratios and chlorophyll concentration.

The most familiar family of such relationships uses a simple idea: take the ratio of reflectance in blue bands to reflectance in a green band, then map that ratio to chlorophyll using an empirically tuned curve. The ocean is bluer when chlorophyll is low and greener when chlorophyll is high, so the ratio changes smoothly across many orders of magnitude. This is the logic behind widely used "OCx" algorithms (often described as "OC4, OC3," and related variants depending on sensor and band set). Their elegance lies in their practicality: a complicated biogeochemical state is reduced to a small number of spectral measurements and a calibrated conversion. Their limitation is equally plain: they work best where the open ocean behaves like the calibration data, and they struggle where other substances—suspended sediments, colored dissolved organic matter from rivers, or unusual plankton communities—distort the optical signal.

A historical anecdote makes the achievement feel less inevitable. In 1978, NASA launched the Coastal Zone Color Scanner (CZCS), a pioneering instrument that was, in some sense, an experiment with a thin budget and uncertain political backing. Many scientists were skeptical that ocean color could be measured reliably from space; the atmosphere seemed too much of a nuisance, and the ocean too dark a target. CZCS operated intermittently into the 1980s and produced the first global glimpses of phytoplankton patterns: productive upwelling zones ringing continents, seasonally shifting blooms at high latitudes, and vast subtropical gyres with persistently low chlorophyll. Those early maps were coarse by modern standards, but they changed the conversation. A phenomenon that had been stitched together from ship transects and scattered stations could suddenly be seen as a planet-scale texture.

That texture is not merely aesthetic. It links the small-scale physiology of phytoplankton to large-scale circulation. The green filaments off Peru and Namibia trace upwelling systems where winds draw deep, nutrient-rich water to the sunlit surface. The springtime brightening of the North Atlantic reflects not only increasing light but the seasonal dance of mixing and stratification. Even the desert-like subtropical gyres carry their own signature: low chlorophyll at the surface, often with an active subsurface layer that satellites largely miss, reminding us that "what you see" is constrained by optics.

That constraint is central. An ocean-color sensor does not see the entire euphotic zone. It sees an *optical depth* weighted toward the surface, typically the upper few to few tens of meters depending on water clarity. In very clear water, photons can emerge from deeper; in turbid water, the sensor's view becomes a thin skin. This matters because the ocean often hides biomass beneath a relatively clear surface, as in the deep chlorophyll maximum common in stratified regions. A satellite map can report low surface chlorophyll while a substantial amount of photosynthetic machinery sits below the depth that contributes much to water-leaving radiance.

Even when the sensor captures the surface layer accurately, chlorophyll itself is an imperfect proxy for carbon biomass or for photosynthetic rate. Phytoplankton adjust their pigment content depending on light and nutrient conditions. In dim environments, cells often pack in more chlorophyll per unit carbon to harvest scarce photons; in bright environments, they can reduce pigment while maintaining growth, partly to avoid light stress. Two water parcels with the same chlorophyll concentration can therefore have different amounts of carbon and different productivity. Ocean color provides a powerful index of the surface ecosystem, but it is not a direct readout of "life" in any single sense.

The desire to turn chlorophyll maps into carbon fixation maps led to another layer of inference: satellite-based primary production models. One influential approach, developed in the 1990s by Michael Behrenfeld and Paul Falkowski, uses surface chlorophyll along with information about available light to estimate depth-integrated photosynthesis. The idea is straightforward in principle: productivity depends on how much photosynthetic machinery is present (approximated by chlorophyll) and how much light it receives, modulated by physiological parameters that represent how efficiently phytoplankton turn light into carbon under prevailing conditions. The challenge is that productivity is a depth-integrated quantity, while satellites observe mainly the surface, and the vertical structure can vary widely.

To bridge that gap, models must assume something about how chlorophyll and photosynthetic capacity are distributed with depth. In some regions, a simple vertical profile might be adequate. In others, especially where deep chlorophyll maxima dominate, a surface measurement can underestimate the integrated stock of pigment and the location of production. Sophisticated models incorporate climatological profiles or use additional satellite products such as diffuse attenuation (a measure of how quickly light decays with depth) to infer how deep light penetrates. Still, the basic limitation remains: satellites view the ocean obliquely, through the atmosphere, and mostly see the skin.

This is where the "algorithmic" nature of the ocean record becomes unavoidable. A satellite-derived chlorophyll time series is not just a history of phytoplankton; it is also a history of sensor calibration, atmospheric correction methods, band choices, and algorithm updates. Operational data products are periodically *reprocessed* as understanding improves. Reprocessing is not a sign of failure; it is a sign that the community treats the record as quantitative rather than merely illustrative. But it has consequences. A trend seen in an older processing version may shift when calibration drifts are corrected or when atmospheric models are improved. The ocean may not have changed; the interpretation did.

Uncertainty enters at multiple points, and it can be easy to forget that a smooth global map hides a great deal of probabilistic machinery. Clouds block the view entirely, forcing gaps that are later filled statistically when producing gridded products. Sun glint can contaminate the signal and must be avoided or corrected. Aerosols vary rapidly in time and space. Coastal waters add complexity because sediments and dissolved organic matter alter reflectance in ways that mimic or mask chlorophyll. Even the act of matching satellite pixels with ship measurements for calibration and validation is tricky: a ship samples a small, moving point, while a satellite pixel averages a much larger area over a short time window, and the ocean is rarely still.

None of this negates the value of the view from orbit. It clarifies what kind of value it offers. Satellites excel at revealing patterns: fronts, eddies, boundary currents, blooms that ignite and spread, and the quiet persistence of oligotrophic deserts. They provide consistent, repeated coverage that ships cannot. A research vessel might spend weeks crossing an ocean basin and still miss an ephemeral bloom by a few days. A satellite can watch that bloom rise and fade, and can do so repeatedly across years.

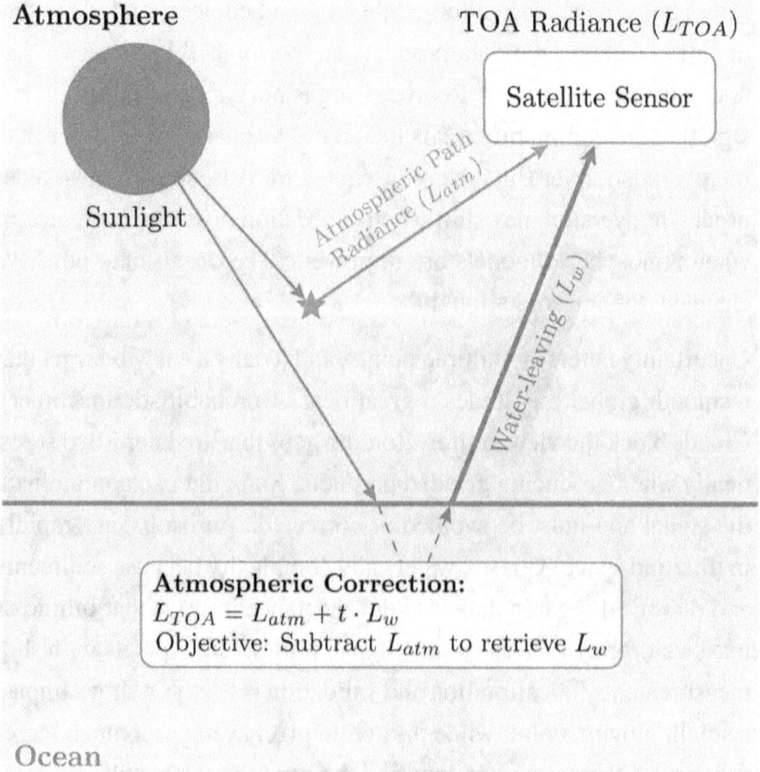

There is also a cultural shift embedded in these maps. They encourage thinking of biology as something that can be tracked like weather. A bloom becomes a moving feature that can be monitored, compared across years, and related to physical drivers such as wind, currents, and heat flux. That framing has real-world stakes. Harmful algal blooms in coastal regions can be monitored and, in some cases, anticipated by combining satellite observations with models and in-water measurements. Fisheries management benefits from knowing where productive fronts and upwelling filaments are likely to form. Climate science depends on estimates of how much carbon the ocean's surface life fixes and how that varies with climate oscillations.

Yet the view from orbit can also tempt a false confidence: the idea that because a map is global, it is complete. The ocean resists that. The subsurface layers shaped by stratification, the deep chlorophyll maxima, the thin layers, the physiology-driven shifts in pigment per cell—these are all ways the sea can hide or distort what color alone suggests. The satellite record is strongest when it is treated as one strand in a braided measurement system: complemented by ships, moorings, floats, and laboratory work that reveal the vertical structure and the physiological state of the organisms responsible for the color.

When that braid is tight, the result is astonishing: a planet-scale observatory for microscopic life. The ocean's surface becomes a living screen on which currents, seasons, and organisms write their signatures in blue and green. A sensor passing overhead in silent orbit cannot see a single cell, cannot tell a diatom from a cyanobacterium, cannot taste nitrate or feel turbulence. But it can detect the collective optical consequence of trillions upon trillions of photosynthetic machines, each negotiating light and survival in a layered sea, and it can do so often enough to turn fleeting blooms into an interpretable record of the breathing ocean.

Chapter 3

The Hunger Games of the Microscopic

Beneath the waves, a trillion-dollar energy economy operates with ruthless efficiency, governed not by currency but by carbon. This chapter descends into the invisible fray where zooplankton commuters, bacterial scavengers, and viral saboteurs battle for resources, ultimately deciding whether the ocean stores carbon or breathes it back into the atmosphere.

3.1 The Billion-Ton Commute

At noon, the open ocean can look like a desert. Sunlight pours down, the surface glints, and if you trail a net through the upper water you may haul up surprisingly little that is large enough to see without a microscope. Then the light begins to soften. The sea is still outwardly calm, but a mass movement starts in the water column: animals the size of rice grains, sesame seeds, and poppy seeds begin to rise. Copepods (tiny crustaceans), krill, amphipods, larvaceans, and a menagerie of other zooplankton leave deeper, darker layers and climb toward the surface to feed. Hours later, before dawn, they reverse direction and sink back down again.

This is diel vertical migration (often shortened to DVM): a daily commuting pattern timed to the light-dark cycle. It is so widespread and so synchronized that it has been called the largest migration on Earth, not because each individual travels far, but because the number of individuals moving is enormous. The distances are modest by human standards—tens to hundreds of metres for many species, sometimes more—yet the consequence is planetary. When the commuters eat near the surface and then *process* that food at depth, they act like living conveyor belts that move carbon downward. In an ocean where the fate of carbon hinges on minutes and micrometres, the simple decision of when to be near the surface can influence how much carbon stays in contact with the atmosphere, and how much is sent into the long, slow interior circulation.

The timing is not a mystery once you adopt the viewpoint of a creature that is both hungry and hunted. The sunlit surface layer is a buffet: phytoplankton grow there because they need light, and phytoplankton are the foundation of most marine food webs. But light is also a spotlight. Many zooplankton are nearly transparent, yet they cast enough of a silhouette that visual predators—small fish, jellyfish with light-sensitive organs, even other zooplankton—hunt more effectively in bright conditions. To be a copepod at noon near the surface is to be surrounded by food and surrounded by danger, a dilemma that plays out, countless times each day, across the global ocean.

Evolution has produced a behavioral compromise. Feed under cover of darkness, when the buffet is still open but the spotlight has dimmed; hide deeper during daylight, where food is scarcer but the risk of being seen is reduced. DVM is not a single fixed pattern. Some species migrate only a little; some stay shallow even by day; some reverse the timing; and some populations switch strategies with season, moon phase, or life stage. But the general logic is consistent: light sets the schedule because light controls both photosynthesis and predation.

If the ocean were a static column of water, this would be merely an ecological curiosity. The commuting matters because of what zooplankton do with what they eat. A copepod grazing on a phytoplankton cell is not only transferring energy upward in the food web; it is also repackaging carbon. It respires some of the carbon as carbon dioxide, it excretes dissolved waste, and it produces fecal pellets—compact, often fast-sinking particles that can drop through the water column much more rapidly than a loose flake of dead plankton. In other words, DVM and zooplankton feeding do not simply move carbon between organisms. They change the *form* of carbon and the *depth* at which that carbon is released back into the water as dissolved CO_2 or as sinking particles.

Oceanographers often describe the ocean's *biological carbon pump* as a set of processes that take carbon fixed by phytoplankton in the surface and transport it to depth, where it can be stored away from the atmosphere for years to centuries. The word "pump" is not metaphorical flourish; it is a recognition that physics alone does not guarantee that carbon leaves the surface layer. Organic matter can sink passively, but much of it is eaten, decomposed, or dissolved before it travels far. A pump needs mechanisms that either speed the trip downward or shift the release of carbon below the well-lit, well-mixed surface. Zooplankton provide both: their pellets speed sinking, and their commuting shifts where respiration and excretion occur.

A useful way to see this is to separate two kinds of export. *Passive export* happens when particles formed near the surface sink under gravity: dead phytoplankton, aggregates, and fecal pellets that begin their descent in the upper ocean. *Active export* happens when animals move carbon downward in their bodies and then release it at depth by respiration, excretion, defecation, or death. The same pellet can be part of either story, depending on where it is produced. A pellet released near the surface joins passive export; a pellet released hundreds of metres deeper has skipped the gauntlet of predators and microbes in the surface layer and begins its journey already closer to the deep ocean.

The biological stakes for the commuter are immediate. The biogeochemical stakes for the planet are cumulative. Carbon released at 500 metres is not the same as carbon released at 50 metres. In the upper ocean, winds and waves mix water back to the surface quickly; carbon dioxide there can equilibrate with the atmosphere on relatively short timescales. Deeper down, water is more isolated, and the return path to the surface is governed by slow circulation and mixing. The commuters exploit that vertical structure without knowing it, the way a human commuter exploits a subway system without understanding the engineering of tunnels and tracks. The ocean has a layered architecture, and behavior turns that architecture into a pathway for carbon.

One of the simplest ways to grasp the scale is to think in terms of flux: how much carbon crosses a depth boundary per day. Primary production in the sunlit ocean is large, but the fraction that escapes downward is often modest because so much is recycled near the surface. If DVM increases export by even a small fraction, the absolute amount can be substantial when integrated over ocean basins. Modeling studies that explicitly include diel migration tend to find a meaningful contribution to export, though the precise number depends on how migration behavior is represented and on local conditions. In some regions and seasons, active transport can rival or exceed passive sinking for certain depth horizons, especially when migrators are abundant and their daily travel spans the boundary between the surface layer and the twilight zone beneath.

The twilight zone—oceanographers call it the mesopelagic—is

roughly the layer from about 100 to 1000 metres. Light there is too dim for photosynthesis, but not absent; it is enough for some animals to see silhouettes, enough for bioluminescence to matter, and enough for a daily rhythm to persist. It is also a region where the chemistry of respiration becomes visible at large scales. When organic matter is respired, oxygen is consumed and carbon dioxide is produced. If a great deal of respiration is relocated into a particular depth range by daily migration, that relocation can help shape oxygen profiles and can contribute to subsurface zones where oxygen is naturally low. In some parts of the ocean, the margin between oxygen-rich and oxygen-poor water is thin. A commuting biomass that concentrates its metabolic demand there can, in effect, thicken that boundary.

This gives DVM an unexpected role: it is not only a response to predators; it can be a driver of habitat. The distribution of oxygen influences where many animals can live, and it influences microbial decomposition and nutrient cycling. If migrators intensify oxygen depletion in certain layers, they may indirectly influence the very predators and prey that made migration advantageous in the first place. The ocean is full of such feedback loops, where behavior and chemistry shape one another over long times.

The commuting pattern is so consistent that humans noticed it before they understood what caused it. A concrete historical thread runs through the development of sonar in the early twentieth century. Engineers and naval scientists, using sound to probe the water column, discovered a puzzling “deep scattering layer”: a band that reflected sonar strongly and that moved up and down each day. During wartime, this was at first a nuisance and a worry. Was it a phantom fleet? Was it a change in water properties? The layer’s daily rhythm made it hard to dismiss, and hard to interpret. Over time, careful observation and net sampling revealed the truth: the scattering was caused by animals—a living layer of zooplankton and small fishes, migrating in unison. In a sense, the ocean announced the existence of the billion-ton commute by interfering with human technology. What seemed like noise in an instrument turned out to be a biological signal of global significance.

That anecdote carries a lesson that remains relevant for climate science. The living ocean is not a decorative skin on a physical system; it is part of the machinery. Instruments designed to measure water can end up measuring life, because life changes the distribution of particles, and particles change the scattering of sound and light. The same is true for carbon. When we measure carbon export, we are not measuring a passive rain of detritus alone; we are measuring the combined outcome of physics, chemistry, and behavior.

Consider a single copepod on its nightly ascent. It rises into warmer water where phytoplankton are abundant. It grazes, filling its gut with carbon-rich material. Some of that carbon is assimilated into tissue; some is turned into eggs; some is burned for energy and returned to the water as CO_2. Then, as dawn approaches, the copepod descends. During the day it continues to respire. It may also defecate. A fecal pellet produced at depth is a tight bundle of organic carbon that may sink rapidly through layers where microbial decomposition would otherwise erode the flux. It is not guaranteed to reach the deep sea—many pellets are intercepted, fragmented, or eaten—but it begins its journey with a head start.

Now multiply that copepod by trillions. The commute becomes a flow of carbon that has a schedule. Nighttime concentrates grazing in the surface layer; daytime concentrates processing at depth. The result is a daily modulation of carbon pathways. The surface is not merely a place where carbon enters the food web; it is also a place where carbon can be retained by rapid recycling. The commuters, by removing carbon at night and releasing part of it deeper by day, alter how tightly the surface retains its carbon.

The details depend on the architecture of the water column. In a strongly stratified ocean—one with a sharp density gradient that resists mixing—the surface layer is thinner and more isolated from depth. That isolation can intensify both the opportunity and the risk for migrators. A thin surface layer can concentrate food and predators, sharpening the day-night contrast. It also means that exporting carbon below the surface layer can more effectively separate it from immediate contact with the atmosphere, because the barrier to upward mixing is stronger. In contrast, in a seasonally mixed ocean where storms deepen the surface layer, the boundary that defines "export" is itself moving. The commuters then interact with a shifting target: a depth that is sometimes within the mixed layer, sometimes below it.

Light itself is also shaped by biology. Phytoplankton pigments absorb light; a productive surface layer can shade what lies beneath. That shading changes how deep the commuters need to go to find safety. A darker ocean can allow migrators to remain shallower by day, reducing the commuting distance and potentially reducing the depth at which carbon is released. A clearer ocean can push them deeper. Here, the feedback is almost cinematic: phytoplankton bloom, the water darkens, and the commuters alter their schedule and depth in response, which then affects how carbon is exported and where oxygen is consumed.

It is tempting to imagine that the commuters always strengthen the biological pump by ferrying carbon downward. But the story is subtler. Active transport can increase the amount of carbon that crosses the boundary of the sunlit layer, yet it can also increase respiration in the twilight zone, converting organic carbon back into dissolved CO_2 at intermediate depths. Whether that CO_2 is effectively sequestered depends on how long the water stays away from the surface and how mixing and circulation distribute it. Carbon respired at 200 metres in a region of strong winter mixing may return to the surface within a season. Carbon respired at 600 metres in a more stratified region may remain isolated longer. The commuters are a pump, but the storage tank depends on physics.

There is also an ecological price to paying for export with pellets. Pellets are valuable food. Many organisms specialize in eating them, breaking them apart, or colonizing them. A fast-sinking pellet can be slowed by being shredded; it can be repackaged into slower-sinking fragments; it can dissolve; it can be turned into dissolved organic matter by microbes. The descent is a gauntlet of consumption. This is why the commuters' depth matters so much: every metre of head start reduces the time pellets spend in the most crowded layers, where interception is most likely.

The daily rhythm also knits together two worlds that feel separate: the sunlit ocean that produces new organic carbon, and the dim interior that must decide what to do with it. The twilight zone is not a passive receiver of whatever sinks down; it is an active marketplace of predation and decomposition. Migrating zooplankton are both customers and couriers. They arrive with carbon in their guts, and they leave behind carbon in their wake, as CO_2, dissolved wastes, and pellets. Predators track them. Microbes track the particles they produce. The commute is a predictable delivery schedule in an ecosystem where predictability is rare.

If one wants an image, it is less a rain of detritus than a timed shipment. Passive sinking is gravity's work: a constant drizzle punctuated by occasional storms of particles. Diel vertical migration is behavioral logistics: a mass of organisms that chooses when to expose itself to danger in order to eat, and chooses where to hide in order to survive. The carbon follows those choices.

From the standpoint of climate, what matters is that this behavior changes the efficiency with which surface production is exported and stored. The ocean currently absorbs a significant fraction of the carbon dioxide released by human activities, buffering atmospheric increase but at the cost of ocean acidification and other changes. The strength and character of the biological pump influence how that uptake is partitioned between surface and depth, and how long carbon remains isolated. If warming increases stratification, if oxygen minimum zones expand, if phytoplankton communities shift toward smaller cells, then the commuters will not simply keep doing what they have always done. Their predators will change, their food will change, and the light environment will change. The commute is ancient, but it is not fixed.

There is a final irony worth holding onto: the commuters are tiny, and their motion is slow. A copepod cannot outswim a current, cannot defy turbulence, cannot control the ocean's heat content or its salinity. Yet by coordinating their risk-taking with the turning of the planet, they have found a way to make their small bodies matter at the scale of the carbon cycle. Each dusk, each dawn, the ocean rearranges its living mass, and the chemistry follows.

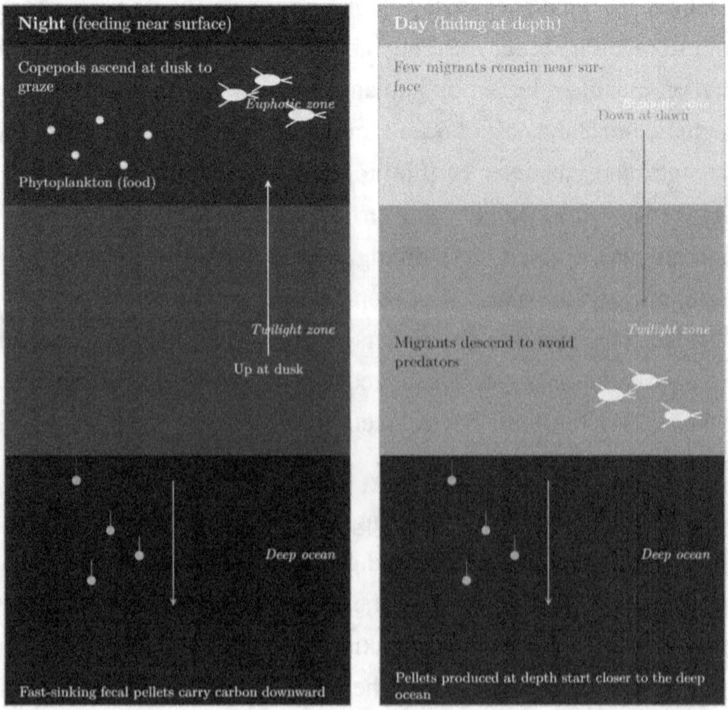

Diel Vertical Migration (DVM): nightly ascent to feed, daytime descent for safety, with carbon exported by respiration, excretion, and sinking fecal pellets.

What makes the commute feel almost algorithmic is its sensitivity to a few cues that are measured, implicitly, by small nervous systems: the brightness of the water, the direction and intensity of light, the shadows cast by particles, the presence of predators sensed through turbulence and chemicals, the hunger level signaled by an empty gut. From those cues emerges a collective pattern that repeats with the reliability of a timetable. It is not perfect, and it does not need to be. In a turbulent ocean, even approximate synchronization is enough to produce a moving layer that instruments can detect and that biogeochemistry must respond to.

One can stand on a ship at dusk and watch the surface darken, then lower a net and recover it glistening with life that was absent hours earlier. The haul is mostly water and translucence, but inside that translucence is a daily decision that has been repeated for millions of years: risk the surface to eat, retreat to depth to survive, and in doing so carry a portion of the sun's carbon away from the air that surrounds the sea.

3.2 Scavengers of the Invisible Soup

A glass of seawater looks clean until you learn how to look at it. Hold it up to the light and it seems like emptiness with a few drifting specks. Yet most of its organic carbon is not in those specks at all. It is dissolved: molecules and fragments of molecules, too small to see, too dilute to smell, and so thoroughly mixed that they feel more like chemistry than like food. To the organisms that live by eating other organisms, dissolved organic matter is an insult—a feast turned into mist. To microbes, it is a landscape.

Dissolved organic matter (DOM) is exactly what it sounds like: carbon-containing compounds dissolved in seawater, along with dissolved nitrogen, phosphorus, and other elements bound up in organic form. When the focus is carbon, oceanographers often say dissolved organic carbon (DOC). The crucial point is that DOC is not a niche pool or an exotic curiosity. It is one of the ocean's largest reservoirs of organic carbon, vastly outweighing the amount contained in the ocean's living biomass at any moment. That mismatch can feel counterintuitive. How can the ocean hold such a large stock of organic carbon without looking like soup? The answer is that it *is* soup, but made of molecules rather than chunks, and a large fraction of it is at concentrations so low that only specialists—enzymatic specialists—can make a living from it.

The story of how that soup is eaten is the microbial loop, a phrase that started as an argument and ended as a foundation. Early pictures of marine food webs often looked tidy: phytoplankton (microscopic algae) use sunlight to build biomass; zooplankton graze them; fish eat the zooplankton; larger predators eat the fish. That picture is not wrong so much as incomplete. It leaves out the most common kind of organism in the sea. Bacteria and other microbes are not decoration at the bottom of the chain. They are accountants, recyclers, and sometimes thieves, intercepting carbon that would otherwise flow upward to animals or downward to the deep.

The microbial loop begins with leakage. Even in a world of efficient grazers, organic carbon constantly slips out of solid form and becomes dissolved. Phytoplankton release part of what they produce as exudates: sugars, amino acids, and other compounds that ooze from cells or are actively secreted. Zooplankton feed messily; they tear, crush, and shred, and the water around their mouthparts becomes enriched with dissolved molecules. Animals excrete. Cells die and burst. Sunlight can also modify organic matter, breaking some molecules into smaller pieces and making others harder to use. The surface ocean, especially when it is warm and well lit, is a factory not only for particles but for dissolved carbon.

Once carbon is dissolved, it might seem as if it has escaped the food web. In fact it has entered a different one. Bacteria and archaea (another group of microbes) take up dissolved organic compounds, using them as both energy and building material. Some of that carbon becomes microbial biomass—new bacterial cells. Some is respired to CO_2. Either way, microbes transform the invisible soup into something with consequences: CO_2 that can return to the atmosphere, and microbial cells that can be eaten.

That last step is the loop. If bacteria only respired dissolved carbon, they would be a sink that short-circuits the transfer of energy to larger organisms. But bacteria are food. They are grazed by small protozoans, including heterotrophic flagellates and ciliates—single-celled predators that swim and hunt in a world where viscosity matters and every encounter is an event. Those protozoans, in turn, are eaten by larger zooplankton. Dissolved carbon is therefore not necessarily lost to the classic food chain; it can be repackaged into microbial biomass and fed back upward, albeit with extra metabolic tolls at each step.

The word *loop* is not just descriptive; it is a reminder that energy and carbon can circulate locally rather than march neatly upward. In much of the open ocean, particularly in nutrient-poor subtropical gyres, that circulation dominates. When nutrients are scarce, phytoplankton are often small, and small phytoplankton tend to leak more dissolved material relative to their size. Grazers may be smaller too, and sloppy feeding is harder to avoid when your prey is tiny and abundant. In those regions the ocean can look blue and empty, but its microbial economy is busy. The loop keeps carbon in the surface system longer, increasing the fraction that is respired near the atmosphere and reducing the fraction that is packaged into sinking particles.

This has direct implications for the biological carbon pump. Carbon export to depth depends on carbon being turned into forms that sink, or being carried down by migrating animals. Dissolved carbon does not sink on its own. It moves with water. If a large share of primary production is rapidly converted into DOC and then respired by bacteria in the surface ocean, less of that production is available to be exported. The microbial loop can therefore act like a local retention system: it keeps carbon circulating in the sunlit layer, powering microbial predators and their predators, while bleeding energy away as heat via respiration.

Yet it would be a mistake to frame microbes as villains in a climate story. They are also the reason the surface ocean does not become clogged with organic debris. They regenerate nutrients that phytoplankton need. They make the ocean habitable for organisms that depend on rapid recycling in low-nutrient settings. And their transformations of DOM can, paradoxically, contribute to long-term storage. Some fraction of dissolved organic matter is resistant to rapid microbial consumption; it persists for years to millennia, and it can be transported into the ocean interior by circulation. The ocean's DOC reservoir contains both fast food and near-inert leftovers, and the balance between these fractions is part chemistry, part ecology, part physics.

The microbial loop became widely accepted because it solved puzzles that the tidy chain could not. One puzzle was numerical: measurements of bacterial abundance and bacterial production in seawater suggested that bacteria were not surviving on scraps; they were processing a substantial share of the ocean's organic carbon. Another puzzle was energetic: protozoans were clearly abundant and active in the water column, but what were they eating if not phytoplankton? Bacteria filled the gap. The conceptual breakthrough is often associated with Farooq Azam and colleagues in the early 1980s, who pushed the idea that microbes are central players in pelagic ecosystems rather than background noise. Their timing mattered. New methods for counting and measuring microbes were emerging, and a generation of oceanographers trained on nets and fish was suddenly confronted with a world where much of the action was happening at scales smaller than their mesh.

A concrete example helps. Imagine a phytoplankton bloom that begins with a nutrient pulse: perhaps a storm mixes deeper water upward, or a current brings nutrients into the sunlit layer. Phytoplankton multiply. Zooplankton respond, but not instantly. In the meantime, the bloom leaks dissolved compounds into the surrounding water. Bacteria, which can reproduce quickly, take advantage of that pulse almost immediately. Their numbers rise, and with them the numbers of their grazers: flagellates and ciliates. By the time the zooplankton population catches up, a substantial fraction of the bloom's carbon may already have been routed through microbial mouths and microbial metabolisms. Some of it becomes CO_2 near the surface. Some becomes microbial biomass that is eaten and turned into new particles. Some becomes dissolved compounds altered by microbial enzymes into forms that linger.

Now place that scene in the context of the daily commuting described earlier. Zooplankton rising at night find a surface ocean in which not all carbon is packaged as cells. A portion is dissolved. Zooplankton cannot graze dissolved molecules directly; they need particles. Microbes, however, can convert dissolved carbon into particulate form by becoming particles themselves: bacterial cells are small particles, and protozoans that eat them are larger ones. In that sense, the microbial loop can *restore* edibility. It turns invisible soup back into bite-sized biomass. But the restoration comes with a cost. Each microbial conversion step typically loses carbon to respiration. The loop is therefore a trade: it recovers some carbon for the larger food web while also consuming a portion of it as metabolic fuel.

The ocean is full of microhabitats where this trade becomes intense. Among the most important are organic aggregates, often called marine snow: clumps of dead cells, mucus, fecal material, and minerals that stick together into flakes and strings that sink. From far away, an aggregate is a particle; from close up, it is a city. Bacteria attach to it, swarm over it, and secrete enzymes that break its polymers into smaller molecules. Those molecules leak back into the water as dissolved organic matter, creating a halo of food that attracts other microbes. Inside an aggregate, gradients can develop: oxygen can be high at the surface and lower inside; nutrients can be concentrated in pockets; acidic microzones can form where respiration is intense. This is not the open ocean as a uniform dilute environment; it is patchy, structured, and sometimes crowded.

Aggregates matter for carbon export because they are major vehicles of sinking flux. They also matter because they are major sites of remineralization—the conversion of organic carbon back to dissolved inorganic carbon and nutrients. Microbes on aggregates can speed that conversion, reducing how much of a particle's carbon reaches depth. But they can also alter particle stickiness and density, influencing whether particles fragment or consolidate. Some particles become faster as they gain ballast from minerals or are repackaged into denser forms. Some become slower as they are eroded. The microbial loop is therefore not confined to the dissolved phase. It interlocks with particle dynamics, deciding whether carbon becomes a sinking ticket to the deep or a dissolved meal that keeps the surface community running.

It is helpful, too, to notice that the loop is not a single pathway. Bacteria differ in what they can eat. Some specialize in simple compounds; others are equipped with enzymes for complex polymers. Communities on particles can change rapidly as early colonizers break down the easiest substrates and later arrivals feed on the byproducts. Even within a few days, an aggregate can host a succession of microbial types, each exploiting a different niche created by the others. The ocean's carbon economy is therefore not only about the amount of carbon in each pool, but about the *rules* of access: who has the enzymes, who arrives first, and who can survive the competition.

The word "algorithmic" is often used loosely in discussions of nature, but here it has a literal flavor. Enzymes are decision rules written in protein. If a bacterium has an enzyme that cleaves a particular bond, it can turn that bond into energy. If it does not, the bond might as well be locked. Microbial communities collectively compute which dissolved compounds are consumed quickly and which accumulate. Over long periods, that computation shapes the composition of the global DOC pool. A molecule's persistence is not just chemistry in isolation; it is chemistry filtered through the distribution of microbial capabilities.

Oceanographers sometimes ask a deceptively simple question: what fraction of primary production is exported out of the surface ocean? The answer is not a fixed constant because export efficiency depends on the community. A food web dominated by larger phytoplankton that form particles readily, and by grazers that produce dense fecal pellets, tends to export more. A food web dominated by small cells and intense microbial recycling tends to export less. The microbial loop sits at the center of that distinction. It is one reason two patches of ocean with similar sunlight can have very different impacts on carbon storage.

A historical vignette gives the loop a human face. In the 1970s and early 1980s, as oceanographers began to take microbial measurements seriously, there was resistance not because the data were wrong, but because the implied picture was uncomfortable. If bacteria and protozoans are central, then much of the ocean's productivity is processed through organisms too small for traditional sampling gear. Nets, long the emblem of biological oceanography, suddenly looked blunt. The shift required new tools: epifluorescence microscopy, radioisotope tracers, flow cytometry. It also required an intellectual shift: accepting that a large share of the ocean's energy flow is routed through organisms that do not have names in field guides and do not appear in children's books about the sea. The microbial loop was, in that sense, a democratization of the ocean: it gave agency to the countless invisible individuals that had always been there.

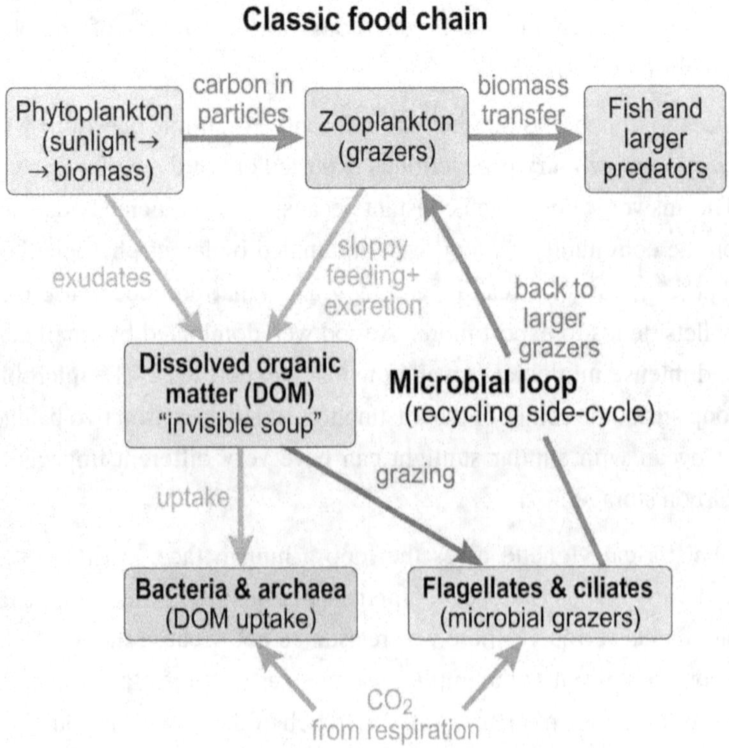

A diagram like this can make the loop look optional, as if the ocean has a main road and a scenic detour. In many waters the detour is the main road. When primary production is dominated by tiny cells, when grazing is diffuse and continuous rather than concentrated in big bites, the dissolved fraction can be a large share of the daily budget. Microbes then determine how much of the day's photosynthesis becomes dinner for animals and how much becomes CO_2 again.

A human analogy can be surprisingly faithful if it stays modest. Think of the surface ocean as a city that imports energy each day in the form of sunlight. The classic chain resembles a supply route that moves goods directly to large consumers. The microbial loop resembles a system of repair shops and recyclers that intercept scraps, fix them, and resell them, while using some of the energy to keep their own lights on. The city can run either way, but the balance changes the amount of waste exported and the amount burned locally.

That balance matters for the living ocean too. Fish production, for example, tends to be higher where carbon moves efficiently from phytoplankton to zooplankton to fish—a shorter chain with fewer respiratory losses. Those are often regions with larger phytoplankton, like productive coastal zones and upwelling regions. In oligotrophic gyres, the microbial loop can dominate, and the ecosystem may support large biomasses of microbes and small grazers but relatively low fish yields. The invisible soup is not only a carbon story; it is a story about which parts of the ocean feel generous to humans and which feel barren, even when the microscopic world is thriving.

One more twist brings the loop back to the carbon pump. Dissolved organic matter does not sink easily, but it can be exported physically when surface waters are subducted or mixed downward. In winter, surface cooling and winds can deepen the mixed layer, pushing surface water into the interior. In some regions, currents peel off layers of surface water and slide them beneath warmer layers, carrying whatever dissolved compounds they contain into the ocean interior. If some fraction of DOC is resistant enough to survive microbial consumption for months to years, it can hitch a ride into deeper waters and contribute to longer-term carbon storage. The microbial loop therefore influences not only *how much* carbon is respired quickly, but also *which* dissolved compounds persist long enough to be transported away from the atmosphere.

The surface ocean can thus be understood as a negotiation between leakage and capture. Every act of photosynthesis creates organic carbon; every act of feeding and every act of cell maintenance risks turning some of that carbon into dissolved scraps. Bacteria turn scraps into biomass and CO_2. Protozoans turn bacterial biomass into larger packages. Zooplankton and their pellets connect the surface to depth. Physical circulation decides which dissolved leftovers are buried in the interior. None of this is tidy, and none of it is wasted. The ocean runs on leakiness. Without it, microbes would starve; with too much of it, export weakens and the atmosphere stays in closer contact with the carbon fixed by plankton.

Seen from the deck of a ship, the loop is invisible. Seen from the perspective of the water itself, it is constant, intimate work: molecules slipping out of cells, enzymes reaching into solution, microbial mouths closing on nanoscopic meals, and predators hunting those mouths. The sea looks empty because the action is mostly dissolved, and the scavengers are mostly too small to cast shadows. Yet the difference between a particle that sinks and a molecule that is respired can be the difference between carbon stored for a week and carbon stored for a century, and that difference is often decided by organisms that fit comfortably on the tip of a needle.

3.3 The Viral Short-Circuit

A virus has no eyes, no gut, no muscles, and no metabolism of its own. It cannot swim toward food or away from danger. It is, in a strict sense, inert until it collides with the right living cell. Yet in the sea it is difficult to take a sip of water without also taking in viruses. Their abundance is so great that they form a kind of biological background radiation: mostly unnoticed, constantly present, and capable of sudden, local catastrophe.

The unsettling part is not merely that viruses kill. Predation kills too, and so does starvation. The unsettling part is *what kind of death* a virus delivers. When a copepod eats a phytoplankton cell, carbon remains packaged in bodies and pellets that can be transferred, reworked, and sometimes exported downward. When a virus infects a microbial cell and causes it to burst, much of that cell's carbon is spilled into the water as dissolved and very small particulate material. The meal becomes mist again. In an ocean where the biological carbon pump depends on packaging carbon into sinkable forms, viral killing can act like a sabotage of packaging: it keeps carbon in the recycling zone rather than letting it move efficiently into larger bodies or into sinking particles.

This is the viral shunt, a phrase coined to capture the way viruses divert organic matter away from the classic grazing pathway and into dissolved organic matter that fuels bacteria. The word *shunt* is revealing. It is not a separate food web so much as an electrical bypass, a short wire that reroutes current. The surface ocean runs on carbon fixed by phytoplankton, and that carbon has several routes: it can be eaten by grazers, exported in pellets and aggregates, or recycled in place. Viral lysis shifts the balance toward recycling. It makes the microbial loop more powerful by feeding it directly with the contents of living cells.

A simple picture helps, as long as it is not taken too literally. Imagine a phytoplankton cell as a tiny warehouse of carbon, nitrogen, and phosphorus, wrapped in a membrane. Grazers are like trucks: they take warehouses whole, move them into larger warehouses, and sometimes drop compact parcels that sink. Viruses are like a lockpick and a demolition charge combined. They enter the warehouse, force it to manufacture more lockpicks, and then blow the doors off so that the goods scatter into the street. In a crowded microbial city, scattered goods are not wasted; they are snapped up quickly by bacteria. But scattered goods are harder to export. They are also harder to keep out of the atmosphere, because their rapid bacterial consumption produces CO_2 in the surface ocean, where mixing and gas exchange keep the water in frequent contact with the air.

The viral shunt is therefore not only a story about mortality. It is a story about the form of carbon. Viruses turn particulate, edible biomass into dissolved and small particulate material, which is more likely to be recycled locally. They also redistribute nutrients in ways that can sustain new microbial growth in the surface layer, sometimes prolonging productivity in a bloom by keeping nutrients circulating among microbes rather than locked up in larger organisms. From the standpoint of a fishery, that can look like inefficiency: energy that might have fed copepods and fish is consumed by bacteria and protozoans instead. From the standpoint of microbial life, it is a bonanza: an episodic release of rich cellular contents into the water, like a sudden spill of sugar and amino acids.

Viruses are not equally important across all organisms. A copepod can be infected by viruses, and there are viruses that infect marine animals, but the dominant ecological and biogeochemical effects in the open ocean come from viruses infecting microbes: bacteria, archaea, and phytoplankton. These hosts are small, abundant, and fast-growing, which creates a steady supply of susceptible cells. Viral infection thrives in a world of high encounter rates and rapid turnover.

It can feel strange to speak of encounter rates in a fluid where everything is mixed, but at microscopic scales the ocean is not a smooth continuum. It is patchy and structured: chemical trails behind swimming cells, nutrient plumes around particles, microzones near aggregates. Viruses drift and diffuse through these landscapes, and their chances of hitting the right host depend on turbulence, viscosity, and the density of potential victims. In a dense bloom, infection can spread like fire through dry grass; in clear oligotrophic water, the same virus may drift for long periods before meeting a suitable cell. This sensitivity to host density gives viral effects a dramatic rhythm. They can be background mortality for long stretches and then, suddenly, become the dominant cause of death.

A concrete example comes from the end of some phytoplankton blooms, especially those dominated by large, distinctive groups. One of the most famous is the coccolithophore *Emiliania huxleyi*, a phytoplankton that covers itself in tiny plates of calcium carbonate called coccoliths. Satellite images can sometimes detect its blooms because the plates scatter light, turning patches of ocean a milky turquoise. Those blooms can be spectacular, and they can collapse rapidly. In many cases, viruses that specifically infect *E. huxleyi* have been implicated as major agents of bloom termination. The collapse is not merely a decline in cell numbers; it is a transformation of the ocean's carbon packaging. As infected cells lyse, dissolved organic matter rises sharply, bacterial activity can surge, and the bloom's carbon is rerouted from particles toward recycling.

That example is scientifically rich because it is also visually and emotionally vivid. A bloom is an apparent triumph of photosynthesis: sunlight turned into biomass at scale. A viral epidemic is a nearly invisible reversal: billions of tiny infections turning that biomass into dissolved fragments. It is a reminder that the ocean's most consequential events are often not storms or whales breaching, but molecular-level transactions happening in plain water.

The historical path to recognizing viruses as major ocean players is itself a cautionary tale about invisibility. For a long time, marine virology was not a central discipline. The sea's viruses were too small for light microscopes, and for decades they were easy to overlook unless one had specialized methods and a reason to look. The turning point came with the rise of electron microscopy and improved counting techniques in the late twentieth century. Once people began to count virus-like particles in seawater, the numbers were impossible to ignore. The ocean was not merely full of bacteria; it was full of their predators and parasites too, on a scale that rivaled or exceeded other biological components.

That recognition forced a conceptual expansion. Predation by protozoans was already a complication to the tidy chain; viruses added another layer. Protozoan grazing generally converts bacterial biomass into larger, more edible packages, potentially allowing carbon to climb back toward zooplankton and fish. Viral lysis does the opposite: it converts biomass into dissolved substrates, strengthening bacterial growth but weakening transfer to larger organisms. The microbial loop and viral shunt therefore form a coupled system. In some conditions the loop can feed higher trophic levels; in other conditions the shunt can keep most of the action among microbes, with repeated cycles of growth and lysis.

Viruses do more than kill cells; they shape who survives. In a community of microbes, the most abundant types may become the most vulnerable because they offer the most targets. This idea has been framed as "killing the winner": viruses can prevent a single microbial type from monopolizing resources by preferentially infecting the most common host. The result is a kind of enforced diversity, an ecological pressure that maintains a mosaic of microbial strategies. That diversity matters for biogeochemistry because different microbes have different metabolic capabilities. If viruses help maintain a broader palette of microbes, they indirectly influence which kinds of dissolved compounds are consumed quickly, which persist, and which transformations dominate.

They also influence evolution more directly. Viruses carry genes, sometimes including genes that alter host metabolism during infection. Some marine cyanobacteria (photosynthetic bacteria) are infected by cyanophages that can carry photosynthesis-related genes, allowing infected cells to keep photosynthetic machinery running during the infection cycle. This is not altruism; it is viral pragmatism. Keeping the host's energy production running can help produce more viral copies before the cell bursts. But the ecological implication is striking: viruses can act as vectors of genetic material and as temporary managers of host metabolism, blurring the line between infection and reprogramming. In a world where microbial communities collectively determine carbon fate, a virus that tweaks metabolism is not a minor actor.

The carbon consequences of viral lysis can be traced through a few linked steps. When a cell lyses, it releases dissolved organic carbon and nutrients. Bacteria take up those dissolved compounds rapidly, increasing bacterial production and respiration. Protozoans may then graze on the bacteria, returning some carbon toward larger sizes, but each step loses some carbon to respiration. The net effect is often to increase the fraction of primary production that is respired in the surface ocean rather than exported. Put bluntly: viruses can turn potential sinking carbon into near-surface CO_2.

There is an important nuance, because the ocean refuses to be simple. Viral lysis can sometimes increase the stickiness of organic matter, promoting aggregation and potentially aiding export—an idea sometimes discussed as a "viral shuttle." If lysed cell contents include sticky polymers, they can help glue particles together. Aggregates can then sink, carrying both organic material and mineral ballast downward. Whether the shunt or the shuttle dominates depends on conditions: the composition of the host community, the chemistry of exudates, the presence of minerals, and the physical environment. But even in the shuttle picture, the mechanism remains the same: viruses change *form*. They decide whether carbon stays as discrete living cells, becomes dissolved soup, or becomes sticky material that clumps.

The daily vertical migration of zooplankton adds yet another layer of interaction. Migrating grazers can reduce host densities near the surface at certain times, altering encounter rates between viruses and phytoplankton. Conversely, if viruses reduce the abundance of certain phytoplankton, they may influence the food available to migrators, potentially changing grazing pressure and fecal pellet production. These feedbacks are difficult to untangle because they operate on different schedules: viruses can act on timescales of hours to days; migration acts daily; seasonal mixing and nutrient supply act over weeks to months. The ocean's carbon pathways are therefore not governed by a single dominant clock, but by multiple clocks that sometimes align and sometimes interfere.

A dinner-party way to capture the difference between grazing and viral lysis is to ask what happens to a loaf of bread. If you cut it into slices and pack it into a lunchbox, you can carry it somewhere else. If you put it in a blender with water, you can still consume it, but you have made it much harder to transport without spilling. Grazing slices and packs carbon; viral lysis blends it into the water. Bacteria are the ones who can drink it. Export, in many cases, depends on carbon staying lunchbox-friendly.

For climate, this matters because the ocean is not simply a sink for carbon dioxide; it is a set of processes that determine *how long* carbon stays away from the atmosphere. Carbon that is converted to CO_2 in the surface mixed layer can equilibrate with the atmosphere relatively quickly. Carbon that is exported to depth as sinking particles or active transport can be stored longer. The viral shunt tends to bias the system toward rapid recycling near the surface, reducing the efficiency of export. In a warming world, if stratification increases and nutrient supply to the surface declines in some regions, microbial communities may shift toward smaller cells and tighter recycling—conditions under which viral interactions can be particularly influential. The precise outcome is uncertain because viruses respond to host communities, and host communities respond to physics and chemistry. But the direction of importance is clear: it is difficult to predict the carbon cycle without taking seriously agents that are smaller than a bacterium.

There is also a human relevance that is not only climatic. Marine microbes, including phytoplankton, produce a large share of the oxygen in the air we breathe and underpin marine food webs that support fisheries. Viral control of microbial communities therefore touches both oxygen production and the transfer of energy to higher trophic levels. In coastal waters, viral lysis can influence harmful algal blooms, sometimes contributing to their collapse and sometimes complicating their dynamics by recycling nutrients locally. The ocean's viral ecology is not merely a theoretical curiosity; it can intersect with water quality, ecosystem health, and the stability of food resources.

Despite their minimalism, viruses are not simple. They come in diverse forms, infect diverse hosts, and participate in diverse interactions. Some infections are lytic, ending in cell bursting. Others are temperate, integrating into host genomes and lying dormant, potentially altering host traits and reactivating under stress. Environmental conditions can tilt the balance. Ultraviolet light can damage viral particles; turbulence can change contact rates; temperature can alter replication speed. Each of these factors adds conditional logic to the viral role in carbon cycling.

It may help, finally, to return to the basic contrast that runs through the surface ocean's carbon economy: export versus recycling. Zooplankton commuting and fecal pellets are among the mechanisms that package and move carbon downward. The microbial loop recovers dissolved scraps and keeps energy circulating, sometimes feeding back to larger grazers but often with significant respiratory loss. Viruses, by lysing cells, pour fuel into that loop and often keep carbon close to the surface by dissolving and fragmenting what could have been eaten or sunk.

In the end, the viral short-circuit is not a moral story of good exporters and bad recyclers. It is a story about control knobs. The ocean's ability to store carbon is not set only by currents and temperature; it is tuned by the countless microscopic interactions that determine whether carbon is packaged, dissolved, clumped, consumed, or respired. A virus is among the smallest levers in nature, but it can push on one of the largest machines on Earth: the balance between the carbon we keep in the sea and the carbon that returns, molecule by molecule, to the air above it.

Chapter 4

The Perfect Recipe

Life in the ocean appears chaotic, yet it adheres to a strict atomic ledger written in the precise ratio of carbon, nitrogen, and phosphorus. When this elemental balance tilts, it determines whether a sea becomes a teeming fishery or a desolate blue desert, revealing how the scarcity of a single atom can drive planetary climate cycles.

4.1 The Golden Ratio of the Sea

If you want to understand why the ocean behaves less like a soup and more like a system with a memory, you could do worse than begin with a piece of numerical poetry: $106 : 16 : 1$. Carbon, nitrogen, phosphorus. Counted as atoms, not as grams. Measured not in one bay or one season, but as a stubborn regularity that seems to echo from the sunlit surface all the way down into the cold interior.

Alfred Redfield was not looking for poetry. He was a physiologist by training, interested in how organisms work, and he had the practical temperament of someone who expects nature to be constrained: by chemistry, by metabolism, by the costs of building living tissue. In the early twentieth century, oceanography was becoming a discipline with its own ships, its own sampling bottles, its own expanding archive of numbers. There were the new chemical measurements of seawater nutrients made possible by improving analytical methods; there were painstaking counts and compositions of plankton collected in nets; and there was a growing awareness that life and chemistry might be entangled in more than a one-way relationship.

The entanglement Redfield noticed was, at first glance, almost too neat. In the upper ocean, tiny drifting organisms take up nitrate and phosphate to build proteins, nucleic acids, membranes, and all the crowded machinery of a cell. Those organisms die, are eaten, or are broken apart; their remains sink as flakes, pellets, and clumps of mucus, the "marine snow" that carries surface chemistry into the dark. Down in deeper waters, bacteria and other microbes consume that material and return much of it to dissolved form: carbon dioxide, nitrate, phosphate. This "remineralization" is, in effect, a set of bookkeeping transactions between living tissue and seawater.

Redfield looked at the books and saw the same signature written in two different hands. The average composition of plankton—what you would get if you could grind up a representative sample and count the atoms—was close to $C : N : P = 106 : 16 : 1$. Meanwhile, the ratio of nitrate to phosphate in deep ocean water was close to $N : P \approx 16 : 1$. The startling part was not simply that these numbers were in the same neighborhood; it was that they were so consistent, across regions and depths, in a medium as vast and apparently variable as the sea.

It is hard, now, to recover the feel of that surprise. Modern biology is saturated with ratios: carbon-to-nitrogen in soils, nitrogen-to-phosphorus in lakes, the elemental signatures of ecosystems mapped from space. But in Redfield's time, the ocean interior was still, in a cultural sense, an unknown place—cold, dark, far away, and thought of primarily as a physical reservoir. To claim that the chemistry of the deep ocean might be shaped by the requirements of drifting microbes near the surface was to reverse an intuitive order. It suggested that life did not merely *tolerate* the ocean's composition; it participated in setting it.

A small but telling historical detail makes Redfield's logic easier to appreciate. The measurements he drew on were not generated by a single grand expedition with a unified plan. They were patchworks: nutrient profiles measured on different cruises, plankton analyses collected by different investigators, the kind of "miscellaneous" data that can look noisy until someone asks the right question. Redfield's gift was not a new instrument but a new comparison. He treated the ocean as a connected system in which the surface and the deep are linked by biological traffic, then asked what that traffic would leave behind if it were large enough, steady enough, and global enough.

That traffic has a direction. Sunlight enters at the top, and that is where photosynthesis happens. Phytoplankton—single-celled algae, cyanobacteria, and other microscopic drifters—use light energy to turn carbon dioxide into organic matter. Yet carbon, by itself, does not make a cell. Proteins require nitrogen; DNA and RNA require both nitrogen and phosphorus; membranes require phosphorus in many organisms; and energy transfer depends on phosphorus-rich molecules such as ATP. If the surface ocean runs short of nitrate or phosphate, it does not matter how much carbon dioxide is available: growth slows, the food web thins, and the export of material to depth declines.

What made Redfield's observation powerful is that it tied this biological need to an ocean-scale chemical pattern. Deep waters accumulate nitrate and phosphate because the organic matter raining down from the surface is continually being taken apart. The deep ocean is, among other things, a long-term compost pile. If the compost is made mostly of plankton whose internal composition tends to hover around a particular $N : P$, then the dissolved nutrients returned by decomposition should reflect that same ratio. In that sense, the deep ocean is a record of averaged life at the surface, smeared out over decades and centuries by mixing and circulation.

It is worth pausing over what a ratio like 16 : 1 actually means, because the seduction of simple numbers can hide their physical meaning. A "molar ratio" counts molecules or atoms rather than mass. Sixteen atoms of nitrogen are not the same weight as one atom of phosphorus; nitrogen is lighter. The ratio is an accounting device for chemical reactions and biological construction. If a phytoplankton cell were a recipe, 16 : 1 would not be "sixteen parts nitrogen by weight," but "sixteen nitrogen atoms for every phosphorus atom" in the mixture of biomolecules that, on average, make up living tissue.

Why should any particular ratio appear at all? Because cells must allocate their limited resources across competing demands. Nitrogen is heavily invested in proteins, including enzymes that harvest nutrients and capture light. Phosphorus, meanwhile, is concentrated in genetic material and in the ribosomes that manufacture proteins. A fast-growing cell tends to require abundant ribosomes and thus relatively more phosphorus; a cell struggling to acquire scarce nutrients may invest in more protein machinery and thus relatively more nitrogen. Even without knowing the biochemical details, you can sense why a universal constant would be unlikely. The ocean contains countless species, each with its own strategies, living in environments that range from the nutrient-poor subtropical gyres to upwelling regions rich enough to stain the sea green.

And yet, when you average across that diversity, patterns emerge—partly because biology is constrained by chemistry, and partly because the ocean itself does a kind of averaging for free. The circulation of seawater mixes parcels from different places; sinking particles blend the remains of many individuals; microbial decomposition integrates over weeks to months; and deep-water formation carries the chemical result into reservoirs that persist for centuries. Redfield's ratio is not a law in the way that the conservation of mass is a law. It is closer to a statistical regularity produced by an immense, continuously running machine.

At the human scale, the key drama is that the ocean is not simply a container of nutrients. It is a container with valves. Consider a concrete example that any coastal community recognizes, even if it does not name it: a spring bloom. As sunlight returns after winter and surface waters begin to stratify (separating into a lighter surface layer and a heavier deep layer), phytoplankton populations can explode. They draw down surface nitrate and phosphate rapidly. Fishers may see the consequences indirectly: changes in water clarity, shifts in zooplankton, the timing of larvae survival. But the chemical trace is unmistakable. A surface layer that was nutrient-rich in winter can become nutrient-depleted in a matter of weeks.

Where do those nutrients go? Some are recycled quickly in the surface by grazing and microbial loops. Some are exported: packaged into sinking particles, carried downward, and remineralized at depth. Over time, the deep water accumulates nutrients; when it is eventually brought back to the surface by mixing, upwelling, or winter convection, it replenishes the inventory that makes the next bloom possible. The ratio at which nitrate and phosphate reappear in deep water is, to a first approximation, the ratio in which plankton used them.

Redfield's insight becomes even more striking when you remember that the deep ocean is not static. Water masses move along the global circulation like slow rivers, descending in high latitudes and eventually returning toward the surface elsewhere. That motion takes time—long enough that the interior ocean behaves like a chemical archive. If the deep nitrate-to-phosphate ratio is consistently near $16 : 1$, then the processes setting it must be persistent and widespread, not local quirks. This is where biology enters as a planetary force. Microbes are not individually powerful, but they are numerous enough to impose their collective stoichiometry on an entire ocean basin.

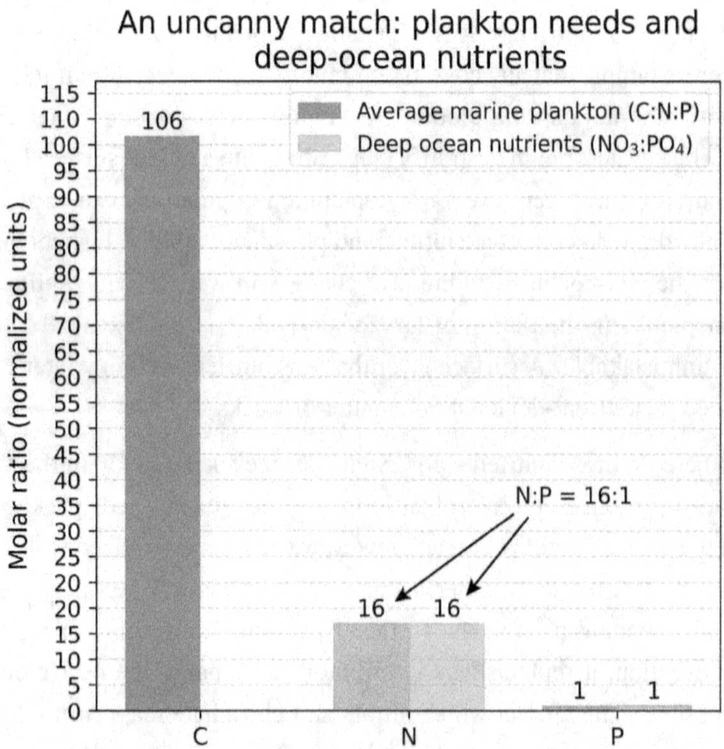

The chart, of course, hides as much as it reveals. Carbon in the ocean is everywhere—dissolved inorganic carbon is abundant compared with nitrate and phosphate—and it participates in buffering reactions that make the chemistry of seawater famously intricate. The point of placing carbon beside nitrogen and phosphorus in the plankton "recipe" is not to claim that deep ocean carbon is simply 106 times something, but to emphasize that biology couples these elements: if you export organic carbon to depth, you usually export nitrogen and phosphorus with it, in a linked package. The deep ocean's nitrate and phosphate, in that sense, are not merely "nutrients waiting to be used"; they are the ash of past photosynthesis.

Redfield's idea was bold enough that it invited argument. Did the biology fix the chemistry, or did the chemistry fix the biology? One can imagine the ocean as a set of constraints imposed from outside: rivers deliver phosphate, the atmosphere provides nitrogen gas, rocks weather, dust falls. Plankton then adapt to what they are given. That story is not wrong, but it is incomplete. Redfield proposed a feedback: if the balance between nitrogen and phosphorus in the ocean drifts away from what plankton, on average, require, then biology itself can nudge it back.

The mechanism involves one of the most ingenious microbial tricks on Earth: nitrogen fixation. Most organisms cannot use the vast reservoir of nitrogen gas (N_2) dissolved in seawater and abundant in the atmosphere, because the bond holding the two nitrogen atoms together is extremely strong. Certain microbes—diazotrophs—can break it, spending energy to convert N_2 into biologically usable forms. If the ocean's fixed nitrogen (nitrate, ammonium) becomes scarce relative to phosphate, nitrogen fixers gain an advantage: they can tap the otherwise inert reservoir and add usable nitrogen to the system. Conversely, if fixed nitrogen becomes abundant, other microbial processes (including those that return nitrogen to N_2 under low-oxygen conditions) can remove it. In this view, the ocean's nitrogen inventory behaves less like a static stock and more like a regulated account, with deposits and withdrawals that respond, imperfectly but meaningfully, to the balance of elements.

Even if you never learn the details of these microbial transactions, the logic of the feedback is intuitive. A cell cannot build proteins without nitrogen or genetic material without phosphorus. If one element is in chronic oversupply relative to the other, it is like having bricks but no mortar: you cannot turn abundance into structure. Any organism that can change the terms of supply—by drawing from a reservoir that others cannot access—will tend to flourish when the balance is off. Over long timescales, that flourishing can help pull the system back toward a characteristic ratio. Redfield's 16 : 1 thus becomes not merely a measurement but a clue about regulation.

There is a danger in taking regulation to mean rigidity. The ocean is full of places where the Redfield ratio is a poor description of what a particular community is doing at a particular moment. In nutrient-poor subtropical regions, plankton often have higher carbon-to-phosphorus ratios than 106 : 1, in part because they are economizing on phosphorus. In upwelling regions, cells may be closer to "balanced growth" and nearer to Redfield. Some taxa, like nitrogen-fixing cyanobacteria, can tilt the local nitrogen budget. Others build mineral shells that affect carbon chemistry and sinking rates. The ocean's stoichiometry is a patchwork, and the patchwork matters because it affects how efficiently carbon is exported to depth and how nutrients are recycled back to the surface.

Redfield's lasting contribution was not to erase that patchwork, but to give oceanographers a baseline—a way to translate between biological production and chemical change. If you know how much nitrate disappeared from the surface, you can estimate, roughly, how much organic carbon might have been produced, and how much phosphate should have been taken up alongside it. If you measure oxygen consumption in the deep sea, you can infer the remineralization of organic matter and predict associated nutrient release. These are not perfect conversions, but they are good enough to turn scattered measurements into coherent budgets, the kind of budgets that reveal whether an ecosystem is being starved, fertilized, or reorganized.

For a popular science reader, the most important implication may be the simplest: the ocean's chemistry is not merely the background against which life happens. It is partly written by life, in a repeating hand, across scales too large for any one organism to perceive. Every phytoplankton cell that divides is making a tiny demand: a little carbon, a little nitrogen, a little phosphorus. Multiply that demand by astronomical numbers, feed the remains into the slow compost of the deep, stir for centuries with currents that connect basins, and a ratio appears—one that you can measure in a bottle of deep seawater hauled up from darkness.

The power of that perspective is emotional as well as intellectual. It asks you to picture the planet's largest habitat not as an inert void between continents, but as an active chemical engine whose settings are tuned, in part, by creatures too small to see. The deep ocean, so often imagined as remote and unchanging, carries in its nutrient proportions a kind of averaged fingerprint of sunlit life. In that fingerprint, $16:1$ is not a magic number. It is the trace left by countless acts of growth and decay, made legible by a physiologist who thought to compare the needs of plankton with the composition of the sea itself.

4.2 The Nitrogen Bottleneck

The ocean has plenty of nitrogen. It is dissolved in seawater and it towers above the sea in the form of atmospheric N_2, a gas that makes up most of the air. Yet marine life regularly behaves as if nitrogen were scarce. Satellite images pick this out as a broad, pale quietness across much of the subtropical ocean: regions where light is abundant but chlorophyll stays low. Out at sea, the scarcity can feel counterintuitive. The water is all around you; the air is all around you; and still the microscopic plants at the base of the food web run on a tight ration.

The trick is that most nitrogen in the world is locked behind a chemical door. Atmospheric nitrogen, N_2, is two nitrogen atoms held together by a triple bond, one of the strongest bonds in ordinary chemistry. For most organisms, that bond might as well be a safe. The nitrogen they can use arrives in "fixed" forms: ammonia, nitrate, nitrite, and a few other compounds that can be slotted into amino acids and nucleic acids. Fixed nitrogen is the spendable currency of marine biology, and the ocean's supply of it is neither infinite nor guaranteed. It is constantly being created, moved, transformed, and destroyed by microbes.

Once you begin thinking in those terms, the surface ocean looks less like a simple pantry and more like a ledger with entries that can be reversed. A phytoplankton cell takes up nitrate to build protein. A grazer eats that cell and excretes waste that bacteria can reuse. A particle sinks and, on the way down, is colonized by microbes that convert organic nitrogen back into ammonium and then, in oxygen-rich waters, into nitrate. These steps recycle nitrogen within the system. But there are also steps that change the size of the account itself: processes that add fixed nitrogen from the vast N_2 reservoir, and processes that return fixed nitrogen back to N_2, removing it from biological circulation.

That add-and-remove tug-of-war matters because nitrogen often becomes the immediate limiter of growth. Phosphorus is essential too, but in large parts of the modern ocean, a parcel of water can contain just enough phosphate to build more biomass if only it had more fixed nitrogen to match it. The stoichiometric "recipe" described earlier gives a useful intuition: if plankton tend to require something like 16 nitrogen atoms per phosphorus atom, then a shortfall in nitrogen relative to phosphorus can stall the conversion of sunlight and carbon dioxide into living matter, no matter how hospitable the temperature and light might be.

The most dramatic form of nitrogen addition is nitrogen fixation: the conversion of N_2 into ammonia by specialized microbes. The protagonists are diazotrophs, organisms that carry the enzyme nitrogenase, a molecular machine capable of breaking the N_2 bond. The process is astonishingly expensive in energy. It is not simply a matter of having the right tool; it is also a matter of paying the electrical bill to run it. This is why nitrogen fixation tends to occur where light is plentiful (to supply energy through photosynthesis), and where fixed nitrogen is scarce enough to make the cost worthwhile. Many diazotrophs are cyanobacteria, close relatives of the organisms that helped oxygenate Earth's atmosphere billions of years ago. In today's ocean they occupy a narrower niche, but a consequential one: they can manufacture the very nutrient that keeps other phytoplankton in check.

A concrete example has been visible to sailors and oceanographers for more than a century, though it is easy to mistake for seaweed or debris. In warm, clear tropical waters, you can sometimes find tawny, floating clumps like damp lint scattered across the surface. This is *Trichodesmium*, a colonial cyanobacterium often nicknamed "sea sawdust". It fixes nitrogen while drifting in sunlit waters that are otherwise poor in nutrients. The colonies are buoyant, so they accumulate at the surface, where a passing ship can scoop them up in a bucket. They have become one of the most recognizable faces of marine nitrogen fixation because they make the invisible visible: a microbial process large enough to tint the sea.

But the ocean is not content merely to add nitrogen. It also subtracts it, and it does so with equal microbial ingenuity. Nitrogen loss happens when microbes use nitrate as an alternative to oxygen during respiration. In oxygen-rich water, organisms "breathe" oxygen, using it to extract energy from organic matter. In oxygen-poor environments, certain microbes can instead use nitrate, stripping away oxygen atoms and ultimately converting fixed nitrogen into N_2 gas. The result is a kind of biochemical unfixing: nitrate that once supported plankton growth becomes inert nitrogen gas again, returning to the vast atmospheric reservoir.

Two related pathways dominate this loss. Denitrification is the stepwise reduction of nitrate to N_2. Anammox (anaerobic ammonium oxidation) is a different route, in which ammonium and nitrite are combined to produce N_2. Both require low oxygen conditions, and both are largely microbial. To an organism living in the surface ocean, these processes are remote. They happen in places where oxygen is scarce: in certain sediments and, most dramatically, in midwater "oxygen minimum zones", layers where oxygen has been consumed faster than it can be replenished.

The existence of oxygen minimum zones is itself a reminder that the ocean's interior is not simply a diluted extension of the surface. In parts of the eastern tropical Pacific, the Arabian Sea, and other regions, winds and currents supply surface waters with nutrients that fuel high productivity. Organic matter sinks, and bacteria consume it as it falls, using oxygen in the process. If the supply of sinking organic matter is large and the ventilation of the subsurface is weak, oxygen can be drawn down to very low levels. These zones are not dead, but they are chemically distinctive, with microbial communities adapted to life on the edge of breath.

Here the nitrogen ledger can flip sign. A parcel of water entering an oxygen minimum zone with a certain amount of nitrate can leave with less. Fixed nitrogen is removed precisely in the places where organic matter is being decomposed most intensely. That linkage creates a feedback with teeth: productive surface waters tend to generate the low-oxygen conditions that encourage nitrogen loss below, which can, in turn, reduce the supply of fixed nitrogen available when those waters return to the surface.

A historical anecdote helps convey how counterintuitive this was when it first came into focus. For a long time, oceanographers treated nitrate as a relatively conservative tracer: something that, aside from biological uptake and remineralization, behaved predictably with circulation. As methods improved and more profiles were collected, it became clear that nitrate could be missing from waters where it "should" have been present, given the phosphate level and the oxygen consumption that signaled remineralization. The mismatch hinted at a sink: a process removing fixed nitrogen beyond simple incorporation into biomass. Unraveling that sink required not only chemistry, but microbiology, and a willingness to accept that major transformations of the ocean's nutrient inventory could be happening in thin layers and dark places.

The Arabian Sea became one of the emblematic arenas for this realization. It hosts a strong oxygen minimum zone, fed by high productivity associated with monsoon-driven dynamics. Measurements of nitrogen gases and isotopic signatures have shown that fixed nitrogen loss there can be intense, and work in the late 2000s highlighted denitrification as a major pathway of N_2 production in that region. The details matter because they change how one balances the ocean's nitrogen budget. If anammox dominates, the requirements for substrates and the sensitivity to environmental change look one way; if denitrification dominates, they look another. The broader point is simpler: the ocean has identifiable regions where fixed nitrogen is actively being converted back into inert gas, thinning the nutrient supply downstream.

This is where the "bottleneck" becomes a planetary story rather than a biochemical curiosity. The surface ocean can behave as nitrogen-limited because fixed nitrogen is being continually removed in oxygen-poor zones and sediments. Yet it does not collapse into permanent nitrogen starvation because nitrogen fixation can, under the right conditions, replenish the account. The balance is dynamic and imperfect. It varies by region, by season, and by the slow changes of climate and circulation. But it is a balance, not a one-way drain.

An accessible way to imagine it is as a two-sided negotiation between microbes with opposed interests. Diazotrophs make fixed nitrogen when it is scarce, but doing so costs energy and requires other resources. Nitrogenase, the key enzyme, is famously sensitive to oxygen; many nitrogen fixers therefore operate in specialized cellular compartments, at specific times of day, or in microenvironments that keep oxygen low enough for the enzyme to function. They also require iron and other trace metals to build nitrogenase, a detail that ties nitrogen fixation to the availability of micronutrients carried by dust and upwelling. Nitrogen loss microbes, on the other hand, thrive where oxygen is low and organic matter is abundant. They do not need to pay the energetic cost of breaking N_2; they make a living by using nitrate as a substitute electron acceptor when oxygen is unavailable.

Neither side "wins" everywhere. In much of the open ocean, oxygen is too abundant for denitrification and anammox in the water column, so nitrogen loss is limited. In many nutrient-poor surface waters, nitrate is too scarce to support high productivity, but the same scarcity can create opportunities for nitrogen fixation. In sediments, where oxygen can be consumed within millimetres of the seafloor, nitrogen loss can be persistent even beneath waters that are well oxygenated above. The spatial patchiness is extreme. Yet the ocean circulation stitches these patches together: water masses formed at the surface carry their chemical signatures into the interior, and the interior ultimately feeds the surface again through mixing and upwelling.

This stitching leads to a deeper question, one that oceanographers have argued over for decades because it is both subtle and consequential: on long timescales, what ultimately controls the ocean's overall productivity? If nitrogen is so often the immediate limiting nutrient at the surface, does that mean the global ocean is ultimately controlled by nitrogen supply? Or is nitrogen limitation a kind of surface-level symptom, while the long-term throttle is set by phosphorus inputs from rocks and rivers, with nitrogen adjusting through fixation and loss to match?

The argument for phosphorus as the long-term controller leans on the existence of nitrogen fixation as a compensating mechanism. If the ocean becomes relatively poor in fixed nitrogen compared with phosphate, nitrogen fixers can add nitrogen until the balance is restored. If fixed nitrogen becomes too abundant, nitrogen loss processes can remove it. In that picture, phosphate behaves more like a slow external constraint: it enters the ocean through weathering and rivers, and leaves primarily through burial in sediments, processes that operate on longer timescales. Fixed nitrogen, by contrast, is more strongly buffered internally by biology. The ocean can therefore exhibit widespread nitrate scarcity at the surface even if, over thousands of years, the total productivity is set by the phosphate budget.

The counterargument is not that phosphorus is unimportant, but that the buffering is neither perfect nor instantaneous, and that regional constraints can persist long enough to shape ecosystems and carbon cycling. Nitrogen fixation is energetically costly and can be limited by iron; nitrogen loss depends on oxygen and organic matter supply and can shift with circulation. The feedbacks may stabilize the global nitrogen inventory, but they can still permit large swings in where and when nitrogen limitation bites. From the perspective of fisheries, harmful algal blooms, coastal hypoxia, and carbon export, those swings are not academic.

A more recent strand of evidence, coming from Earth system models as well as observational syntheses, suggests that the ocean's nitrogen ledger is indeed responsive: adding large amounts of external fixed nitrogen does not necessarily translate into a proportional increase in global marine productivity because enhanced nitrogen availability can stimulate conditions that accelerate nitrogen loss. The details depend on where the nitrogen enters, how oxygen responds, and how circulation distributes the perturbation. In some places the ecosystem response can be strong and immediate, while the global average budges only slightly. This is regulation of a sort, but it is the regulation of a living, spatially structured system, not a thermostat in a single room.

The human relevance becomes sharpest near coasts. Rivers deliver nitrogen and phosphorus from agriculture, sewage, and industry. In estuaries and shelves, where water residence times are shorter and mixing is constrained, the added nutrients can fuel intense blooms that later decay, consuming oxygen and creating seasonal dead zones. These are modern, visible manifestations of the same coupling between productivity, oxygen, and nitrogen transformation that operates in the open ocean. The difference is that coastal systems are more directly connected to human decisions, and their feedbacks can be overwhelmed by sustained nutrient loading. A low-oxygen bottom layer in summer is not just a local tragedy for fish and benthic animals; it is also a chemical environment that favors nitrogen loss, changing the nutrient balance exported back to the broader sea.

It is tempting to search for a single moral in this complexity, but the ocean resists simple morals. Nitrogen is a bottleneck not because it is rare in an absolute sense, but because biology can only use certain chemical forms, and microbes have evolved ways to toggle those forms in response to oxygen, energy, and opportunity. The same element can be abundant and inaccessible, plentiful and limiting, conserved and erased, depending on which organisms are present and what the water allows them to do.

One of the most unsettling implications is that the nitrogen cycle is sensitive to oxygen, and oxygen is sensitive to warming. Warmer water holds less dissolved oxygen; stronger stratification can reduce the ventilation of subsurface layers; changes in productivity can alter the rain of organic matter that fuels oxygen consumption. Oxygen minimum zones are therefore not just curiosities on a map. They are places where small shifts in physics can tip microbial metabolism toward pathways that remove fixed nitrogen. When that happens, the consequences can travel: a change in nitrogen loss in one region can alter nutrient ratios carried by currents, reshaping what limits growth elsewhere.

In the end, the tug-of-war is not a battle between good and bad microbes. It is a kind of distributed governance carried out by organisms with no foresight, responding to local conditions in ways that nonetheless scale up. Diazotrophs, breaking into the nitrogen safe, are doing it for themselves, not for the ecosystem. Denitrifiers and anammox bacteria, returning nitrate to inert gas, are not sabotaging productivity; they are exploiting a niche where oxygen is scarce. The ocean's nitrogen balance emerges from their competing metabolisms, filtered through circulation, seasonality, and the geometry of the seafloor.

A glass of seawater from the surface can look simple: clear, slightly salty, unremarkable. But dissolved in it is a story of chemical access and denial. The surface community may be waiting on nitrogen that is everywhere and yet unavailable, while somewhere below, in a layer too dark for photosynthesis, microbes are quietly converting yesterday's nitrate back into gas. Between those two realms, nitrogen is always in transit: carried by currents, packaged into sinking particles, reshaped by enzymes, and balanced—never perfectly, but persistently—by life that has learned to make and unmake the same molecule for different reasons.

4.3 Give Me Half a Tanker of Iron

There are parts of the ocean where the chemistry reads like an invitation to abundance. Nitrate is high. Phosphate is high. The surface waters have been supplied, by upwelling or winter mixing, with the same macronutrients that drive spectacular blooms elsewhere. And yet the water stays oddly spare. The chlorophyll signal that satellites use as a rough proxy for plant biomass remains low; the familiar greenish tint of a productive sea never quite arrives. These are the High Nutrient, Low Chlorophyll regions, usually shortened to HNLC, and their existence once felt like an insult to common sense. If the pantry is full, why does no one cook?

The answer is that the pantry is not full in the way a cell experiences fullness. The ocean's macronutrients are like flour and rice: essential and bulky, but useless without smaller ingredients that make metabolism possible. One of those ingredients is iron, a trace metal present in seawater at vanishingly low concentrations. Iron is a micronutrient: needed in tiny amounts, lethal in excess, and indispensable precisely because it sits at the heart of key biochemical machines. It helps shuttle electrons during photosynthesis, it supports respiration, and for nitrogen-fixing microbes it is part of the nitrogenase enzyme that breaks the strong bond of N_2. A plankton cell can have all the nitrate and phosphate it wants and still stall, because the internal machinery that turns those nutrients into living matter cannot be assembled without a few atoms of iron in the right places.

The scarcity is not simply a matter of low supply. In oxygenated seawater, iron is chemically fussy. Much of it is present in forms that are not easily taken up, or it is quickly scavenged and removed by particles. The ocean does receive iron from several routes: dust blown off deserts, volcanic ash, sediments stirred up on continental shelves, hydrothermal vents along mid-ocean ridges. But the delivery is patchy and episodic, and the chemistry of iron makes it easy to lose. The result is an ecology that is sensitive to the smallest dribbles of metal, a biological world in which nanomolar changes can move entire food webs.

The existence of HNLC regions became clearer as ocean measurements accumulated across the twentieth century. By the time chemical oceanography had matured enough to map nitrate and phosphate across basins, it was obvious that large areas of the Southern Ocean, the equatorial Pacific, and the subarctic North Pacific contained surface waters rich in these nutrients. Yet the surface phytoplankton were not consuming them down to low levels, the way they do in many temperate coastal seas. The mismatch stood out because it violated the intuitive link between "nutrients present" and "plants grow." Something was preventing the algae from converting macronutrient abundance into biomass and, crucially, into export of carbon to depth.

A sharp idea arrived with a sharp personality. John Martin, a marine chemist, argued in the late 1980s and early 1990s that iron scarcity could be the missing brake in these regions. His proposal became known as the Iron Hypothesis: if iron limits phytoplankton growth over vast areas, then variations in iron supply over Earth's history could have altered ocean productivity enough to influence atmospheric carbon dioxide and climate. During ice ages, the continents were drier and dustier; winds transported more mineral dust to the ocean; the extra iron could have fertilized HNLC regions, enhancing blooms and increasing the ocean's uptake of CO_2. When the climate warmed and dust diminished, iron supply would fall, productivity would ease back, and CO_2 would rise.

It is an idea that feels almost too bold, because it seems to turn something minute into something global. Dust to climate, via microbes. But boldness is not the same as carelessness. The hypothesis had a plausible physical route (dust delivery), a plausible biological mechanism (iron limitation), and a plausible climate link (the ocean's role as a carbon sink). What it lacked, at the time, was a direct test: an experiment big enough to perturb the ocean at a scale where ecology and physics could be measured in the open sea, not inferred from bottles or lab cultures.

Testing it required an unusual kind of fieldwork. You cannot fertilize an ocean basin, but you can fertilize a patch of ocean and watch what happens before currents and mixing erase the signal. Mesoscale iron enrichment experiments were designed to do exactly that: add a trace amount of soluble iron to a defined area, track that water mass with drifting buoys and chemical tracers, and measure the biological and chemical response over days to weeks.

One of the best-known of these experiments, IronEx II, took place in 1995 in the equatorial Pacific. The setting mattered. The equatorial Pacific is a region where upwelling brings nutrients to the surface, but phytoplankton growth often remains modest relative to the macronutrient supply. The experimenters released iron into a patch of surface water and tagged the patch so it could be followed. Then they watched, as the ocean responded.

The response was dramatic in the way only a nutrient-limited system can be dramatic. Phytoplankton biomass rose sharply inside the fertilized patch. Chlorophyll increased. Rates of photosynthesis increased. Nutrients began to decline. Measurements of the carbonate system, which tracks dissolved CO_2 and related chemistry, showed that surface CO_2 inside the patch dropped measurably relative to surrounding waters. It was as close to a controlled "before and after" as the open ocean allows. The result made a simple point with unusual force: iron can indeed be the missing ingredient that unlocks macronutrient use and stimulates carbon drawdown, at least over the short term.

For anyone tempted to hear in that result a straightforward solution to climate change, the ocean quickly supplied complications. The first complication is physical. A fertilized patch does not sit still. It stretches, filaments, and mixes with surrounding waters. Even if you could keep adding iron, you are not fertilizing a static pond; you are perturbing a moving fluid that disperses your intervention. The second complication is ecological. A bloom is not just a pile of carbon. It is a reorganization of a food web. Grazers respond. Some phytoplankton taxa outcompete others. The efficiency with which carbon ends up exported to depth depends on what kinds of organisms dominate and how the food web packages organic matter into sinking forms. The third complication is chemical and physiological. Adding iron can shift not only how much phytoplankton grow, but also their elemental ratios and their demand for other micronutrients. A system can be iron-limited and, once iron is supplied, become limited by something else or by the rate at which light and mixing allow growth.

These complications are not technical footnotes. They determine whether iron fertilization could ever meaningfully increase the long-term sequestration of carbon in the deep ocean, rather than simply creating a temporary surface drawdown that is later undone. The goal, if one is thinking climatically, is not merely to remove CO_2 from surface water for a few days. It is to transfer carbon into the ocean interior and keep it there long enough to matter for atmospheric levels. That requires export, and export requires particles that sink, escape rapid recycling, and reach depths where the return path to the atmosphere is slow.

Here the earlier stoichiometric "recipe" re-enters the story in a more complicated form. If you add iron to an HNLC region and phytoplankton begin consuming nitrate and phosphate, you might imagine a near-theoretical maximum: use up all the macronutrients, produce a great deal of organic carbon in roughly Redfield proportions, export much of it, and thereby draw down atmospheric CO_2. Real blooms rarely reach that ideal. Grazing can recycle nutrients in the surface, keeping them in circulation rather than exporting them. Mixing can dilute the patch with nutrient-rich water from outside, sustaining productivity but also making the net "effect" harder to attribute. The bloom can shift toward smaller cells that are more easily retained and recycled in the surface rather than sinking. The patch can run out of iron again, or run into limitation by light or silica (for diatoms), or by other trace elements. The result observed in some experiments is that the measured CO_2 drawdown can be a fraction of what a naive stoichiometric calculation would suggest if all nitrate were converted into exportable carbon.

This is one reason the Iron Hypothesis remains both compelling and controversial. The experiments support a key premise: iron limitation is real and powerful. Yet translating that into a predictable, climate-relevant lever is far from straightforward. The ocean does not reward single-cause thinking.

The Southern Ocean sharpens the stakes. It is one of the largest HNLC regions and a major gateway between the surface and the deep ocean, strongly connected to global circulation. It is also a region with intense winds, deep mixing, and a harsh seasonal cycle of light. Those physical constraints already make it difficult for phytoplankton to build large standing stocks, even when nutrients are available. Iron supply, delivered partly by dust and partly by upwelling and shelf sources, can still be limiting, but it is limiting within a setting that is physically reluctant to hold onto a bloom.

Southern Ocean fertilization experiments, such as those in the SOIREE era, demonstrated that tiny increases in iron can stimulate blooms and alter nutrient uptake and CO_2 chemistry. These results strengthened the link between iron supply and the biological pump, the process by which the surface ocean exports organic carbon to depth. They also revealed how quickly the system can respond, and how the response depends on community composition and grazing. The Southern Ocean is not merely "iron-starved"; it is a place where biology is negotiating simultaneously with physics, light, and loss.

At the dinner-party level, the temptation is to ask: if iron fertilization makes blooms, why not do it deliberately? Why not add iron to the ocean to help draw down CO_2 and cool the planet?

The first answer is that "making a bloom" is not the same as "sequestering carbon." A bloom can be a surface spectacle with little long-term effect if most of the carbon is quickly respired back to CO_2 and returned to the atmosphere as the fertilized waters equilibrate with air. The second answer is that the ocean is an ecosystem, not a carbon capture machine, and changing one limiting nutrient can have cascading effects. Some phytoplankton produce toxins; some blooms can alter oxygen levels when they decay; shifts in the base of the food web can propagate to fisheries and biodiversity. The third answer is governance. The ocean is shared, and deliberate, large-scale interventions raise questions about consent, liability, monitoring, and unintended consequences that do not have easy technical fixes.

Even leaving policy aside, the scientific caution is grounded in a simple ecological truth: when you add a limiting nutrient, you are not merely feeding "plants." You are selecting winners. The winners may be diatoms that sink efficiently, enhancing export, or they may be small cells that are readily grazed and recycled. They may be organisms that change the balance of nutrients in ways that influence other processes, including nitrogen fixation. Because iron is required for nitrogenase, adding iron can, under the right circumstances, stimulate nitrogen fixation, altering the nitrogen budget and potentially changing which nutrient becomes limiting next. That feedback can be stabilizing in some contexts and destabilizing in others. The system you perturb is not passive.

There is also a subtler climate link embedded in the iron story. The ocean's role in absorbing heat and carbon depends on how surface waters exchange with the interior. Changes in stratification and circulation can alter nutrient resupply and the regions where iron inputs are effective. In a warming world, if stratification strengthens, some surface regions may become more nutrient-depleted, making them less responsive even to added iron because the macronutrients are no longer present. Conversely, changes in wind patterns and dust transport could alter natural iron fertilization rates. The Iron Hypothesis is, at heart, a hypothesis about coupling: between continents and ocean, between physics and biology, between the smallness of micronutrients and the largeness of climate.

A final anecdote captures the peculiar feel of iron limitation: its dependence on invisibly small quantities. In laboratory cultures, an experimenter can cause a phytoplankton population to languish simply by omitting a trace metal or by allowing iron to precipitate out of solution. In the open ocean, the "dose" that matters is not a spoonful but a whisper. It is measured in billionths of a mole. And yet, because phytoplankton are so numerous and because their collective photosynthesis is so large, that whisper can be amplified into a bloom visible from space. Few other environmental processes offer such a clean demonstration of scale: a constraint so small that it can be missed in a contaminated bottle, and so large that it can alter the color of an ocean.

This is the unsettling lesson and the enduring fascination. HNLC regions are not barren because they lack the major ingredients of life. They are barren because the ocean runs on machinery that demands rare parts, and because those parts arrive according to the messy logistics of dust storms, currents, shelf sediments, and volcanic chance.

When iron arrives, whether through a natural pulse of dust or a deliberate release from a research vessel, it can flip a switch that has been holding back growth. What happens after the switch flips depends on everything else: grazers, mixing, light, and the complicated ways that carbon and nutrients are routed through a living web. In the chemistry of a trace metal, the ocean reveals both its sensitivity and its unwillingness to be simplified.

Chapter 5

The Falling Snow of Carbon

Even as surface waters teem with microscopic life, a silent, vertical migration of matter creates one of Earth's most critical thermostats. This chapter traces the journey of biological carbon from the sunlit upper layers into the crushing darkness of the abyss, revealing how the ocean locks away atmospheric greenhouse gases for millennia.

5.1 The Green Engine of the Abyss

The ocean is often described as a carbon sponge, a vast body of water that soaks up some of the carbon dioxide we release into the air. That image is true as far as it goes, but it hides the more interesting fact: a great deal of the ocean's carbon uptake is not a slow, passive chemical process. It is powered. It is built, molecule by molecule, by living cells.

At the sunlit surface, carbon dioxide from the atmosphere slips into seawater and becomes part of a restless chemical crowd: dissolved CO_2, carbonic acid, bicarbonate, carbonate. These forms interconvert constantly, and they matter for ocean acidity, corals, shells, and climate in the long run. Yet none of that chemistry, by itself, guarantees that carbon will leave the surface ocean for any meaningful length of time. Much of the dissolved carbon near the surface can, sooner or later, make its way back to the air.

What changes the story is biology. Phytoplankton—microscopic, photosynthetic organisms drifting in the upper ocean—perform a trick that no current or equilibrium can achieve alone: they take carbon that is available near the surface and package it into the stuff of bodies. Sugars. Proteins. Lipids. The carbon is no longer merely dissolved; it is *organized.* Once carbon is built into organic matter, gravity enters the conversation. A living surface can become a sinking particle. And a sinking particle can carry carbon away from the atmosphere's reach.

That living conveyor belt is the biological carbon pump. It does not mean all carbon is pumped downward, or even most of it. It means that life at the surface creates a one-way bias, a preference for depth, that would otherwise be far weaker. The pump is not a single pipe; it is a mesh of pathways. Some carbon descends as crumbs of dead cells. Some descends as fecal pellets produced by grazers. Some is transported by animals that migrate up and down each day, feeding near the surface at night and releasing carbon at depth. Some is carried downward when surface waters themselves sink or mix. But the central act—the engine that makes the rest possible—is the conversion of inorganic carbon into organic matter by phytoplankton in the thin, bright skin of the sea.

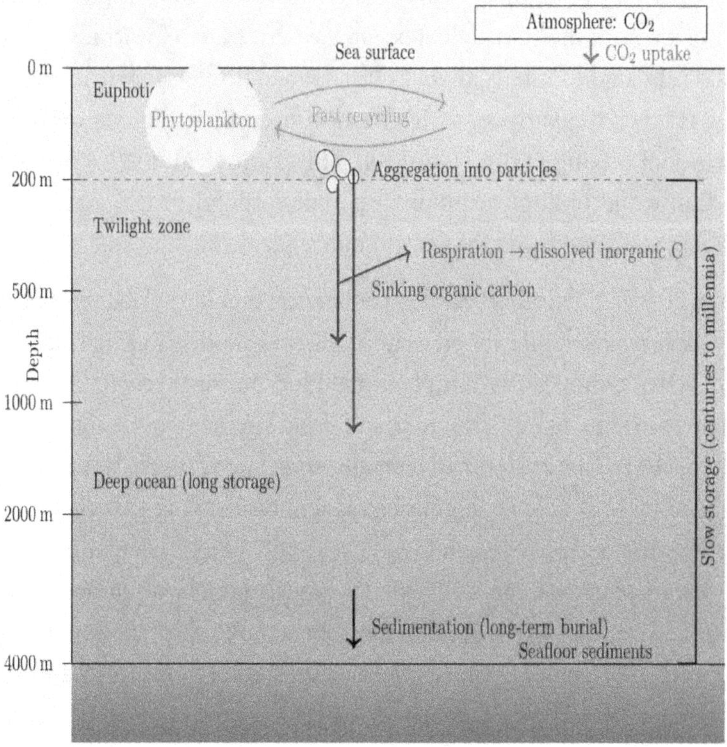

The engine starts with a constraint that shapes almost everything: light fades quickly in water. A few tens of metres down, the world is dim; by a few hundred metres, it is night. Photosynthesis therefore has a fixed address: the euphotic zone, the upper layer where light is enough for phytoplankton to make a living. It is a global habitat, but it is thin. In terms of volume, it is a skin. In terms of consequence, it is a lever.

Phytoplankton are not a single thing. They include tiny cyanobacteria, only a micron or two across, as well as larger cells like diatoms, whose glassy shells can be ornate enough to look like architecture under a microscope. Each group has its own appetite for nutrients and its own way of dealing with scarcity. Some do well in the nutrient-poor gyres of the subtropical oceans, where water is warm, stratified, and reluctant to mix. Others flourish when deeper, nutrient-rich water reaches the surface, as happens in upwelling regions or during seasonal mixing at high latitudes. This ecological diversity matters because the biological pump does not respond to the atmosphere alone; it responds to physics and nutrients as much as to carbon.

A useful distinction is between production that the surface ocean can sustain by recycling its own nutrients and production that depends on a fresh supply. When phytoplankton die or are eaten, most of their nutrients are quickly released back into the surrounding water by bacteria and grazers. In a stable, sunlit layer, those nutrients are like coins that keep changing hands without leaving the room. This is why the open ocean can look deceptively empty: a small standing stock of plankton can still process large amounts of carbon if it is turned over fast. But recycled nutrients tend to keep carbon cycling near the surface. For carbon to be exported downward in a lasting way, the surface often needs an external nutrient subsidy: nitrate mixed up from below, iron delivered by dust, or nutrients introduced by upwelling. That link between nutrient supply and export helped clarify why the biological pump is uneven across the planet, intense in some regions and surprisingly weak in others.

The biological pump can be described in numbers, but it is best understood as a set of pressures. On the one hand, the surface ocean is a place where carbon is easy to acquire, because the atmosphere keeps replenishing it and because the carbonate system holds a large inventory of dissolved inorganic carbon. On the other hand, it is a place where carbon is hard to keep, because warm water holds less gas, because winds and waves continually renew the surface, and because any carbon respired back to CO_2 is only a short diffusion distance from the air. The pump is the ocean's way of making carbon *heavy.*

Consider a bloom: a rapid increase in phytoplankton biomass, often visible from satellites as a spreading stain of green. Blooms can be triggered by seasonal mixing that brings nitrate into the light, by the retreat of sea ice that exposes surface waters to sun, or by pulses of iron fertilization in high-nutrient low-chlorophyll regions. The details vary, but the common feature is speed. When conditions are right, phytoplankton divide at a rate that makes the ocean's surface behave like a fast bioreactor. Carbon is drawn down into organic matter, and oxygen is released, briefly tipping local chemistry toward a more breathable, more basic state.

This is the glamorous part of the story, the part that lends itself to photographs and colour maps. Yet the pump's effectiveness is decided less by how quickly carbon is fixed than by what happens next. The surface is not a factory that ships all its products; it is a market where most goods are consumed immediately. Zooplankton graze on phytoplankton. Bacteria attack dissolved compounds that leak from cells. Viruses rupture plankton and turn living biomass into a soup of fragments. Much of this organic carbon is respired back into CO_2 within days, sometimes within hours. The atmosphere, waiting above, is never far away.

The pump succeeds only to the extent that some of this organic matter escapes the surface recycling loop and enters the ocean interior. The key word is *export*: carbon leaving the surface layer. Export can happen in several ways, but the most intuitive is sinking. A single phytoplankton cell is generally too small and buoyant to fall quickly. In the stillness of the sea, it drifts more than it drops. It becomes export-worthy when it is packaged into larger particles: clumps of dead cells, mucus, fragments of grazed material, and mineral grains that have become stuck together. These aggregates have a different relationship with gravity. They are the first physical expression of the pump: carbon transformed into a particle with a sinking speed.

There is a historical reason scientists sometimes talk about this as a conveyor belt rather than a gentle drizzle. In the mid-twentieth century, the deep sea was still widely imagined as a slow realm, a place of scarcity where little changed. One of the surprises of early deep-ocean sampling—including the use of sediment traps lowered on cables, and later, drifting arrays that could be recovered and analysed—was how dynamic the vertical link could be. Instruments deployed at depth would sometimes return coated with fresh, fluffy material: a mixture of organic debris and microorganisms that looked, under the microscope, like a snowfall of life. The deep sea, it turned out, is fed from above, and much of its diet arrives as particles large enough to be caught.

Those measurements mattered for climate thinking, not only for ecology. If carbon is respired near the surface, it is quickly available to exchange with the atmosphere. If it is respired a few hundred metres down, it is still in water that can return to the surface on decadal timescales in many regions. If it is carried into the deep ocean and respired there, it can be isolated from the atmosphere for centuries. If a small fraction is buried in sediments, it can be stored for millennia or longer. Depth, in other words, is a clock.

The biological pump is therefore less a question of *how much* carbon the surface fixes than of *where* the fixed carbon is ultimately respired. This is one reason the pump has such leverage over atmospheric CO_2. A relatively small change in the average depth at which sinking organic carbon is broken down can change the amount of carbon the deep ocean can store, and thus the amount left in the air. The surface and the abyss are connected by a narrow but consequential accounting rule: carbon that reaches depth tends to stay out of the atmosphere for longer.

It may seem strange to attribute planetary influence to organisms that are individually invisible, but phytoplankton are not rare. In many parts of the ocean, a teaspoon of seawater holds tens of thousands of photosynthetic cells. They are small because the physics of their world rewards smallness: a high surface-area-to-volume ratio helps them absorb dissolved nutrients, and their buoyancy helps them remain in the light. Their small size also means their lives are short and their generations rapid. This gives the biological pump a distinctive rhythm: it is seasonal, weather-sensitive, and capable of responding to sudden changes in temperature and mixing.

The pump is also exquisitely sensitive to the ocean's vertical structure. In warm regions, surface waters are often strongly stratified, sitting like a lighter lid on top of colder, denser water. Stratification limits the upward supply of nutrients from depth, often throttling phytoplankton growth even when light is abundant. In colder regions, or in winter storms, mixing can deepen the surface layer, pulling nutrients up but also sometimes diluting phytoplankton into darker water where photosynthesis slows. The same physical process can therefore both feed and starve the pump, depending on timing. Spring blooms in temperate oceans are, in part, an outcome of this seasonal choreography: nutrients replenished by winter mixing, then a sudden improvement in light and stability that allows phytoplankton to stay near the surface long enough to multiply.

A concrete example comes from the North Atlantic, where the spring bloom has been observed for well over a century and has served as a natural laboratory for understanding export. When conditions align, surface waters can shift from relatively clear to visibly green over large areas. The bloom is quickly followed by a rise in grazers and bacteria that process much of the new biomass. Yet some portion aggregates and sinks, and this pulse of export can be strong enough to be detected far below, arriving as a brief enrichment that feeds deep communities. The timing matters to fish and to fisheries, but it also matters to carbon accounting: the fraction exported during such events helps determine how much carbon is transferred to depth before it can be returned to the atmosphere.

This is also where the biological pump begins to look less like a simple vertical pipe and more like an algorithm: a set of conditional rules executed by a system of countless agents. If the surface is nutrient-rich and calm, phytoplankton may bloom, and grazers may lag, allowing biomass to build. If grazers are abundant early, they may keep phytoplankton in check, converting plant carbon into fecal pellets that sink faster than individual cells. If viruses sweep through a bloom, they may shatter cells into dissolved compounds that bacteria can recycle quickly, keeping carbon near the surface. If dust delivers iron, it can shift which species dominate, changing not only how much carbon is fixed but how easily it forms sinking particles. The pump's efficiency is therefore an emergent property: not a single rate constant, but the result of interactions.

One of the most consequential interactions involves minerals. Many sinking particles are not pure biology; they are composites of organic matter and inorganic ballast. Diatoms contribute silica. Coccolithophores contribute calcium carbonate plates. Dust contributes clays and other minerals. The presence of heavier material can increase sinking speeds and protect organic matter from rapid microbial attack by making aggregates denser and, sometimes, less accessible. This is not a guarantee of deep export, but it is one reason the pump varies by region and by plankton community. A bloom dominated by large, mineral-bearing plankton can export carbon differently from a bloom dominated by tiny cells that are readily recycled.

The biological pump also creates a deep ocean signature that chemists can read. As sinking organic matter is respired at depth, it consumes oxygen and releases dissolved inorganic carbon and nutrients. This helps explain why deep waters are richer in nitrate and phosphate than surface waters and why oxygen levels tend to decrease below the surface in many regions. The ocean interior is, in a sense, where the receipts are stored: a chemical ledger of past surface productivity and of the pathways by which that productivity was processed. Even without watching particles fall, one can infer the pump's long-term activity from the vertical patterns of oxygen, carbon, and nutrients.

Yet for all its global importance, the pump is not a benevolent machine designed to tidy away our emissions. It has limits and vulnerabilities. Warming tends to strengthen stratification in many regions, potentially reducing nutrient supply to the surface and changing the composition of plankton communities. Ocean acidification alters the carbonate chemistry that many organisms rely on for shells and plates. Changes in wind patterns can shift upwelling zones and alter the seasonal timing of blooms. Each of these changes can adjust not only how much carbon is fixed at the surface, but the more delicate question of how much is delivered downward before it is respired.

That delicacy is the core emotional fact of the biological pump: it is simultaneously immense and fragile. Immense, because it is global, continuous, and driven by photosynthesis over an area larger than all land plants combined. Fragile, because it depends on the behaviour of tiny organisms and on the fine structure of the upper ocean—a structure that climate change is already rearranging.

To picture the pump as an engine is not to imagine pistons and gears. It is to recognize that the ocean's surface is not merely a boundary where physics and chemistry meet the air. It is a living membrane that actively reorganizes carbon into forms that can fall. Every sinking particle is a physical message from sunlight to darkness: carbon once in the air, briefly made into life, now passing into the deep where time runs differently.

5.2 The Twilight Blizzard

A single phytoplankton cell is a marvel of chemistry, but it is a poor projectile. In the sunlit surface ocean, it drifts in a world where viscosity matters more than inertia, where a grain of dust can hang in the water like smoke. If the biological carbon pump depended on individual cells simply letting go and falling, the deep ocean would be starved. The carbon that leaves the surface in a meaningful way usually does so after a change of form: from solitary specks into clumps, pellets, and flakes large enough for gravity to win the argument.

That transformation produces something with a misleadingly gentle name: marine snow. The phrase sounds poetic, and divers sometimes describe it that way when they swim through a slow storm of white flecks. In scientific terms, marine snow is a catch-all for sinking aggregates of organic debris and inorganic fragments—dead cells, mucus, tiny shells, dust, fecal material, and the bacteria that colonize it all—stuck together into particles visible to the naked eye. It is not uniform. Some aggregates are fluffy and fragile, like wet tissue paper. Others are compact and fast, especially when packed into fecal pellets by grazers. Some are rich in minerals and sink like grit; others are mostly carbon and water, buoyant enough to hover until they collide with something else.

Marine snow matters because it turns the surface ocean's frantic recycling into a vertical delivery system. It also matters because it mostly *fails*. Most of the carbon packaged into sinking particles is consumed and converted back into dissolved carbon dioxide long before it reaches the seabed. The twilight zone—the dim layer below the reach of photosynthesis, roughly from the base of the euphotic zone down to about a kilometre—is where the majority of that consumption happens. The twilight zone is not empty water between the surface and the abyss. It is an active filter, crowded with small animals, microbes, and chemical reactions that decide how much carbon will be stored at depth and how much will be returned to the fast surface cycle.

If the surface ocean is where carbon becomes heavy, the twilight zone is where that heaviness is tested.

The first step in the test is aggregation. How do tiny particles find each other in a moving fluid? Part of the answer is simple physics. Even in still water, particles do not fall at the same speed. Some sink slightly faster because they are larger or denser. As they descend, they overtake slower particles, collide, and sometimes stick. Turbulence adds opportunities: shear and swirling motions bring particles into contact that would otherwise pass by. The ocean surface, stirred by wind and waves, is therefore not only a place where life grows; it is a place where debris is mechanically introduced to itself.

The second part of the answer is stickiness, and here biology becomes unmistakable. Many plankton produce extracellular polymers—gel-like sugars and other compounds that leak, ooze, or are actively exuded. Under the microscope these substances can look like transparent nets or smears, but in the ocean they behave like glue. They bind cells into flocs, capture dust, and coat tiny mineral grains, encouraging them to clump into larger, sinkable packages. When a bloom collapses, the water can become rich in these adhesives, and aggregation can accelerate dramatically. The ocean's carbon export is often pulsed rather than steady, because aggregation itself can switch on quickly when the conditions are right.

There is also a more intimate kind of packaging: grazing. Zooplankton are not merely consumers; they are particle engineers. When a copepod or krill filters phytoplankton from the water, it compresses diffuse biomass into pellets that can sink far faster than the original cells. A cell that would have hovered near the surface can, after being eaten, become part of a streamlined parcel dropping tens to hundreds of metres per day. From the perspective of carbon sequestration, this is a darkly efficient trade: the surface ecosystem keeps itself alive by eating, and in the process it builds the very vehicles that move carbon out of the surface.

The vehicles, however, are not sealed. A sinking particle is not a lifeless rock; it is a moving habitat. Within minutes to hours, bacteria attach to it and begin to feed. Small protozoans graze on those bacteria. Larger animals bite off pieces or shred aggregates in the course of feeding. Enzymes leak into the surrounding water, dissolving organic matter into smaller molecules that can be taken up directly by microbes. A particle that begins its descent as a coherent flake may be gradually eroded into a trail of dissolved compounds, an invisible wake of carbon that no longer sinks.

This is the twilight zone's central drama: sinking competes with decay. The balance depends on two rates that are easy to say and hard to predict. How fast does the particle fall? How fast is it consumed? If sinking is rapid and decay is slow, more carbon penetrates to depth. If sinking is slow and decay is rapid, the carbon is recycled at relatively shallow levels. Both rates vary enormously, even within the same ocean region, because particles differ in size, density, and composition, and because the organisms feeding on them respond to temperature, oxygen, and the availability of food.

The result is a steep attrition curve. The flux of sinking organic carbon decreases sharply with depth, especially in the first few hundred metres below the surface. Oceanographers sometimes treat this as a kind of law of the water column: a regular decline, predictable enough to put into models, but with parameters that shift from place to place. The practical meaning is straightforward. If you could weigh the falling carbon at, say, 100 metres, and then again at 500 metres, you would almost always find that much has vanished along the way. The twilight zone consumes the snowfall.

The reasons for this consumption are not only microbial. Many twilight-zone animals are specialists in interception. Some drift and wait; others migrate daily. Diel vertical migration—a mass movement of animals that rise toward the surface at night to feed and descend during the day to avoid predators—is among the largest animal movements on Earth by biomass. It is also a carbon transport mechanism. Migrators can ingest carbon near the surface and then respire it at depth, or release it as fecal pellets and dissolved excretions deeper than where it was acquired. In effect, some carbon bypasses the slow physics of sinking particles and is carried by living bodies. This is not marine snow in the narrow sense, but it reshapes the same budget: how much carbon is transferred out of the surface layer, and how deep it reaches before it returns to dissolved form.

To understand why marine snow is so variable, it helps to picture two contrasting particles. The first is a fluffy aggregate formed from diatom detritus and mucus after a bloom. It may be centimetres across but mostly water, with delicate structure and low density. It sinks, but not especially fast. Its large surface area is a feast for bacteria. It may also be easily fragmented by turbulence or by the nibbling of small animals. It begins as a dramatic-looking flake and ends as a dispersed cloud of dissolved compounds.

The second is a fecal pellet produced by a grazer feeding on that same bloom. The pellet is compact, denser, and often coated in a protective membrane. It can fall rapidly, sometimes fast enough to outrun some of the microbial degradation that would otherwise occur. But it is also attractive to pellet-feeders, animals that specialize in intercepting and consuming these parcels. Even fast vehicles can be hijacked.

Minerals complicate the comparison. When organic matter becomes associated with denser materials—calcium carbonate from coccolithophores, silica from diatoms, clays from dust—the particle's sinking speed can increase, and its internal structure can become more resistant to rapid breakdown. The idea of mineral "ballast" is not that minerals magically preserve carbon, but that they change the physical context: a denser particle falls faster and may spend less time in the most biologically active layers, while a more mineral-rich matrix can make some organic matter less accessible. In many regions, carbonate-bearing material has been observed to correlate with deeper carbon flux, hinting that certain kinds of plankton communities may export carbon more efficiently than others. The ocean is not only a biological system; it is a system of mixed materials, and the chemistry of shells can influence the fate of soft tissue.

Measuring these processes at sea has always required a blend of ingenuity and humility. One of the classic tools is the sediment trap: a collector deployed at depth, designed to catch sinking particles over days to months. Early traps were often moored and fixed in place, which made them vulnerable to currents that could tilt them or carry lateral debris into the funnel. Later designs tried to drift with the water to reduce these biases. Even then, traps are not perfect: some swimmers enter and leave, some particles dissolve, some are resuspended. Yet the basic experience of opening a trap cup after a deployment—finding it lined with dark, crumbly material from far above—has repeatedly confirmed that the ocean interior is fed by a rain of particles whose quantity and composition change with seasons, storms, and blooms.

A historical anecdote captures both the wonder and the frustration. In the 1980s, during a period when oceanographers were trying to quantify carbon flux more systematically, large field programs deployed arrays of traps at multiple depths across open-ocean stations. The goal was deceptively simple: measure how much particulate organic carbon leaves the surface and how quickly it diminishes with depth. The surprise was not that the flux declined—that was expected—but how sensitive the decline was to circumstances that were difficult to observe directly. Two stations in the Pacific, for example, could show markedly different transfer efficiency between similar depth horizons: in one case only a modest fraction of the material leaving the upper layer would remain a few hundred metres below; in another, roughly half might persist. Those differences may sound technical, but their implications are stark. If the ocean's twilight-zone filter varies that much across the globe, then global estimates of carbon sequestration can swing by petagrams per year depending on which assumptions are used. The falling snow is not a uniform drizzle; it is a patchy storm system with local weather.

The patchiness is partly ecological. Different plankton communities create different kinds of debris. Small cells tend to be recycled rapidly and may contribute more to dissolved organic carbon than to large sinking particles. Larger cells, especially those that form chains or colonies, may aggregate more readily. Zooplankton community structure matters: a region dominated by grazers that produce large pellets may export carbon differently from a region where grazing is dominated by smaller organisms that shred particles into slower-sinking fragments. Even within a single bloom, the character of export can change over time, as early growth, grazing pressure, and viral lysis alter the mix of particles available to aggregate.

The patchiness is also physical. Turbulence can enhance collisions and aggregation, but it can also break apart delicate flocs. Stratification can keep particles within a thin layer where they are repeatedly processed by grazers and microbes. The depth of the mixed layer—the surface layer stirred into relative uniformity by wind—can determine where newly formed particles begin their descent. A particle released from the base of a deep mixed layer may already be hundreds of metres down, effectively getting a head start on sequestration time, while a particle released from a shallow layer remains in closer contact with the atmosphere's timescale.

Temperature adds a further bias. Microbial respiration generally speeds up in warmer water, which tends to increase the rate at which particles are degraded near the surface in tropical and subtropical regions. Colder waters can slow microbial activity, potentially allowing a larger fraction of export to penetrate deeper. This is not a simple latitudinal rule—ecology and minerals can override it—but it is a reminder that climate change can alter the pump not only by changing how much carbon is fixed at the surface, but by changing the metabolic tempo of the organisms that intercept it below.

One of the most counterintuitive features of marine snow is that it can create micro-environments that are chemically different from the surrounding seawater. Within a sinking aggregate, oxygen can be consumed rapidly enough that the interior becomes low-oxygen even when the surrounding water is well-oxygenated. Nutrients and dissolved carbon can build up in the pores. pH can shift. The aggregate becomes a tiny, moving chemical reactor. For microbes, this is an opportunity: a concentrated patch of food in an otherwise dilute world. For carbon export, it is a vulnerability: intense local respiration can convert particulate carbon into dissolved forms that no longer sink.

The twilight zone is therefore both a marketplace and a battlefield. Gravity pushes downward. Biology pushes toward consumption. The outcome is often described with a single curve, but it emerges from countless encounters: particles colliding and sticking, pellets being produced and eaten, aggregates forming and fragmenting, microbes attaching and digesting, migrators carrying carbon in their guts and tissues. The word "blizzard" is helpful not because the ocean is cold, but because it evokes a three-dimensional storm of particles, each with its own trajectory and fate.

For human beings, the relevance of this hidden snowfall can feel remote until it is framed as time. Carbon that remains in the surface cycle can influence atmospheric composition on the timescales of seasons to decades. Carbon that is respired at a kilometre depth may be held away from the atmosphere for much longer, depending on ocean circulation. Carbon that is buried in sediments may be stored beyond any human planning horizon. The twilight zone controls the handoff between these clocks. It decides whether carbon is briefly delayed or meaningfully sequestered.

There is also an ethical edge to the story, because the twilight zone is increasingly discussed as a place of resource extraction. Interest in mesopelagic fisheries and deep-sea mining has grown, and with it the temptation to treat the midwater as an underused frontier. Yet the twilight zone is not merely a habitat. It is part of the planet's carbon regulation system, a biological filter whose functioning is still only partly understood. Disturbing it without understanding its role is like modifying the insulation in a house because the attic looks empty.

None of this requires viewing the ocean as a fragile ornament. Marine snow has been falling for as long as there has been photosynthesis in the sea. What is striking is the precision with which small changes can alter the balance. A shift in plankton community composition can change particle size and stickiness. A change in grazing can change how much carbon is packaged into fast-sinking pellets. A change in temperature can speed up the microbial communities that dissolve the snowfall. The carbon pump is built from microscopic decisions, and the twilight zone is where those decisions become climate-relevant.

A flake of marine snow begins as a local event: a dead cell, a mucus strand, a speck of dust, a bite taken from a bloom. As it sinks, it becomes a communal object, colonized and reshaped by organisms that never see sunlight. Most of it will be eaten, atom by atom, and exhaled as dissolved carbon in the dim water. The part that survives does not do so heroically. It survives because, for a time, sinking is faster than appetite.

5.3 The Deep Ocean Ledger

Far below the last traces of sunlight, the ocean keeps accounts.

Not in tidy columns, of course, and not with numbers that any one instrument can read directly. The deep sea records carbon as chemistry: a little less oxygen here, a little more dissolved inorganic carbon there; nutrients accumulating in waters that have not seen the surface for centuries; shells dissolving or surviving depending on how corrosive the water has become. For scientists trying to understand Earth's carbon budget—how carbon is partitioned among air, land, and sea, and how long it stays in each place—the deep ocean is both vault and notebook. It stores carbon out of reach of the atmosphere, and it also preserves clues about how that storage has changed through time.

The most immediate problem is deceptively simple: how much of the organic carbon produced near the surface actually makes it into the ocean interior? The biological pump begins with photosynthesis, but sequestration begins only when carbon is carried down and kept there long enough to matter. That is why the deep ocean ledger is less about the existence of sinking particles than about their *fate*. How much carbon falls past a few hundred metres? How much reaches a kilometre? How much arrives at the seabed? And how much is returned to the water column as dissolved carbon dioxide along the way?

One classic way to answer those questions is to try to catch the particles in the act of falling. A sediment trap is essentially a collector deployed at depth, often for weeks to months, designed to intercept sinking material. The simplest version is a funnel leading into a bottle. The best versions are more like an array of carefully designed funnels, baffles, and sample cups that rotate on a schedule so the material can be separated by time. Many are deployed as moorings: anchored to the seafloor, held upright by floats, and positioned at specific depths like rungs on a very long ladder.

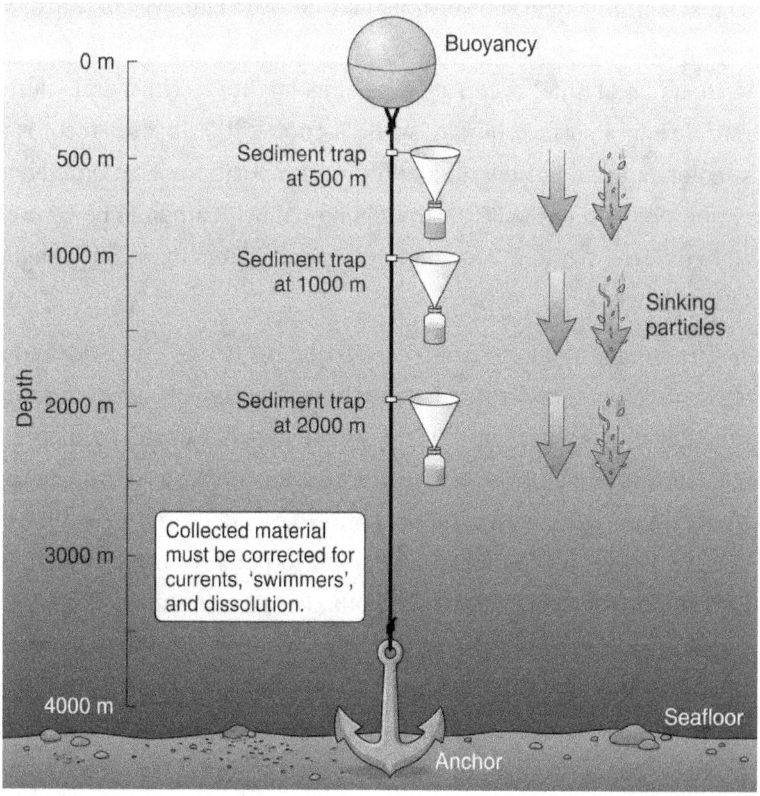

That last sentence in the diagram's box is the reason sediment traps inspire both affection and argument. The ocean is not a still cylinder of water. Currents can tilt a trap, changing what it collects. Some animals actively swim into the funnels (and sometimes out again), contaminating samples with carbon that did not arrive by sinking. Particles can be resuspended from below or carried laterally from elsewhere. Even when a trap behaves perfectly, it measures only what passes through a small area during a particular time. Turning those catches into a global number requires care, cross-checks, and an acceptance that uncertainty is not a failure but a feature of the problem.

The scientific payoff, however, is hard to overstate. When traps are deployed at multiple depths, they show the deep ocean ledger being written in real time. A cup at 500 metres might fill with a dark, fluffy rain during a surface bloom, while a cup at 2000 metres records a smaller, delayed pulse with a different composition. That difference is not merely a curiosity. It reveals how much carbon is being consumed in the twilight zone and how much is making it into longer-term storage.

A particularly memorable moment in the history of this work came with the large, coordinated field programs of the late twentieth century that treated particle flux as something that could be mapped and compared across ocean regions. John Martin and colleagues' VERTEX program in the 1980s, for example, deployed free-drifting sediment traps in the northeast Pacific to quantify how particulate organic carbon diminished with depth. The measurements supported a practical, empirical relationship: flux tends to decline with depth in a way that can be approximated by a power law. That relationship is sometimes treated as a kind of default setting in ocean models, a simple curve standing in for a tangle of microbial and ecological processes. Its usefulness lies not in its perfection, but in the fact that it captures a robust truth: most sinking carbon is intercepted before it can become deep storage.

Yet trapping is only one way to read the ledger, and it is not always the most convenient. There is another strategy that oceanography has made into an art: using naturally occurring chemical tracers as clocks and tally marks. One widely used example is thorium-234, a radioactive isotope produced continuously from uranium dissolved in seawater. Thorium is particle-reactive: it tends to stick to particles and be removed as they sink. If one measures a deficit of thorium-234 in the surface relative to what uranium production would supply, that deficit can be converted into an estimate of particle export over timescales of weeks. In practice, this provides a powerful complement to sediment traps: a tracer-based export estimate that does not require leaving a device in the water for months, and that responds quickly to events like blooms.

These tracer approaches also highlight a subtle point that matters for climate: exporting carbon out of the surface layer is not the same as sequestering it. Carbon that sinks to 150 metres and is respired there may be kept away from the atmosphere for a while, but in many regions that water can return to the surface on relatively short timescales. The deeper the remineralization—the depth at which sinking organic matter is converted back into dissolved inorganic carbon by respiration and decay—the longer the carbon tends to remain isolated from the atmosphere. That depth is a lever. In a modelling study by Eun Young Kwon and François Primeau (2009), deepening the characteristic depth of remineralization by only a few tens of metres on average produced a substantial reduction in atmospheric CO_2 in the model world. The lesson is not that the ocean can be tuned like a thermostat, but that small shifts in where respiration happens can move carbon between reservoirs with very different turnover times.

Once carbon reaches the deep ocean as dissolved inorganic carbon, it joins waters whose movements are slow enough to feel geological. Deep waters are formed in particular places—notably in high-latitude regions where cooling and brine rejection during sea-ice formation make water dense enough to sink—and then they spread through the ocean basins along pathways guided by seafloor topography and the planet's rotation. Oceanographers call the time since a parcel of water last contacted the atmosphere its *ventilation age*. This concept is central to the ledger, because ventilation age is essentially "time in storage." A carbon atom respired into a deep water mass with a long ventilation age is, in practical terms, tucked away.

How does one measure that age? Radiocarbon is one answer. Carbon-14 is produced in the upper atmosphere and enters the ocean through gas exchange. Once in the ocean interior, its radioactive decay becomes a clock. By measuring the radiocarbon content of dissolved inorganic carbon, scientists can estimate how long it has been since that water was last at the surface. The picture that emerges is uneven: some deep waters are relatively young, ventilated on centennial timescales, while others are much older. The deep ocean ledger is not a single vault; it is a set of compartments with different locks and different keys.

The ledger has another remarkable feature: it keeps old records.

Sediment accumulating on the seafloor is not just mud. In many places it is a layered archive of shells, organic fragments, and minerals, laid down year after year. Extracting that archive is the work of coring: pushing a long tube into the seabed and pulling up a column of sediment with the layers preserved. A core is a physical timeline. The deeper you go, the older the material tends to be, though the conversion from depth to age requires dating methods and careful interpretation.

Within those layers are microfossils: the shells of foraminifera and other plankton that lived near the surface or at depth, each building calcium carbonate structures that incorporate chemical signatures of the seawater they grew in. By measuring ratios of stable isotopes (such as $\delta^{13}C$ and $\delta^{18}O$) and trace elements, scientists can infer changes in ocean circulation, nutrient distribution, and carbon storage across glacial cycles. These are not direct readings of atmospheric CO_2, but they constrain the ocean's role in controlling it. The deep ocean ledger, in this sense, is also a memory of climate.

One particularly elegant line of evidence comes from boron isotopes measured in deep-sea corals, which can serve as a proxy for seawater pH. Over the last tens of thousands of years, shifts in deep-ocean pH reflect shifts in the balance between dissolved inorganic carbon and alkalinity, and therefore in how much carbon the deep ocean is holding. Records of this kind help connect the chemistry of the abyss to the familiar story told by ice cores: atmospheric CO_2 rising out of glacial periods and falling into them. When deep waters become more carbon-rich, their pH tends to drop; when they release carbon and become less loaded, pH tends to rise. It is the ledger speaking in a different dialect.

The hard part is turning these many dialects into a coherent budget. A modern carbon budget has to reconcile multiple streams of evidence: direct measurements of sinking particles; tracer-based estimates of export; oxygen and nutrient profiles that reveal cumulative respiration; radiocarbon and other tracers that indicate circulation and storage times; and sediment records that show how the system has behaved under different climates. Each method has its blind spots. Sediment traps sample a small area and can be biased by flow and swimmers. Tracers rely on assumptions about particle affinity and steady production. Chemical budgets in the water column must distinguish between processes that alter oxygen and those that do not. Paleoproxies must be calibrated and interpreted with respect to ecological shifts and diagenesis, the slow alteration of sediments after they are deposited.

What makes the science compelling is that these methods do not merely repeat the same story; they constrain one another. A trap might suggest a certain flux at 1000 metres, while thorium-based export suggests what left the surface, and oxygen consumption at mid-depth suggests how much must have been respired along the way. If the numbers do not balance, something has been missed: perhaps lateral transport, perhaps an unaccounted pathway such as active transport by migrating animals, perhaps an instrumental bias. The ocean forces honesty by being too large and too interconnected to be explained by a single technique.

There is also an emotional charge in the work that is easy to miss from the outside. Deploying a deep-ocean mooring is not like installing a weather station. It is an operation that can fail spectacularly. A storm can ruin the schedule. A line can tangle. An anchor release can malfunction, leaving months of data on the seafloor. Even when recovery works, the samples can be disconcertingly small: a few grams of material representing vast volumes of water. And yet those grams are the deep sea made tangible. They contain the remains of surface blooms, the mineral dust of distant deserts, the signature of organisms that lived and died without ever encountering land.

When people talk about the ocean absorbing our carbon emissions, it is tempting to imagine a single number, a single service being provided. The deep ocean ledger reveals something stranger and more consequential: the service depends on where the ocean chooses to place the carbon in time. Carbon held in the surface ocean is carbon still in conversation with the sky. Carbon carried into deep waters that are slow to ventilate is carbon placed into long storage. Carbon buried in sediments is carbon moved into the planet's geological savings account, mostly inaccessible without extraordinary disturbance.

The falling snow of carbon is therefore not merely a vertical movement. It is an act of bookkeeping, carried out by microbes, zooplankton, minerals, and currents, with the deep ocean holding the receipts.

Chapter 6

The Planet's Lungs

The ocean inhales and exhales on a massive scale, acting as a metabolic partner to the atmosphere while silently buffering our carbon footprint. This chemical intimacy comes at a steep cost, altering the fundamental acidity of seawater and slowly shrinking the habitable volume for complex life.

6.1 The Breath of the Waves

Stand at the shore on a windy day and you can feel the ocean working: spindrift lifted from wave crests, a hiss of foam collapsing, the steady shove of air on your face. It is easy to imagine the sea as a passive basin that merely *contains* water, waiting patiently for the sky to drop gases into it. The truth is less polite. The ocean takes and gives gases through a restless, mechanically driven exchange that depends on wind, temperature, and turbulence. The surface is not a simple boundary; it is a thin, agitated membrane where physics decides how quickly the planet can breathe.

The first surprise is how thin the action is. Most of the ocean is far from the atmosphere in any practical sense: a deep interior connected to the surface only by currents and mixing. The actual gateway is a skin, often far thinner than a human hair in its controlling layer, where molecules must cross from air to water or back again. Scientists call it the *air–sea boundary layer*. The boundary layer is not a physical film you could peel away; it is a region where motion slows, gradients sharpen, and the world changes character. Above it, the air is a turbulent fluid. Below it, the ocean's surface is also turbulent. But right at the interface, things get complicated: the two fluids meet, and for a moment the chaotic swirls of both sides give way to a narrow zone where transfer is limited by molecular diffusion, the slow shuffling of molecules from high concentration to low concentration.

If that narrow zone were the whole story, gas exchange would be a quiet affair. Diffusion is steady but slow, like a faint smell spreading through a still room. Yet the ocean does not breathe like a still room. It breathes like a lung in motion, driven by gusts, waves, and breaking crests that repeatedly tear down and rebuild the near-surface gradients. Wind does not simply carry gases to the ocean; it changes the ocean's ability to accept them.

A useful way to think about this is with the idea of a *transfer velocity*. Imagine there is a difference between how much carbon dioxide the atmosphere wants to push into the surface and how much the surface wants to give back; a similar difference exists for oxygen, nitrogen, and other gases. The net flux depends on that difference *and* on how efficiently the boundary can move molecules across. Transfer velocity is a compact, practical measure of that efficiency. It bundles up an unruly mix of processes—tiny ripples, shear at the surface, short-lived gusts, bubble plumes, even rain—into a single number that answers a concrete question: if the air and the water disagree about a gas, how fast can the argument be resolved?

The ocean's "breathing rate" rises dramatically with wind. In calm conditions, the surface can be deceptively smooth, and the boundary layer on the water side can become thick in molecular terms. Gases still cross, but slowly. As wind increases, it does two things at once.

It makes the air more turbulent, renewing the air-side concentration near the surface, and it makes the water-side surface more turbulent, thinning the diffusive layer that most often limits exchange for many gases. In moderate winds, this thinning is the main effect: turbulence in the top few millimetres and centimetres continually stirs water from below into the surface, preventing a stagnant skin from forming. In strong winds, a second effect takes over: waves break.

Breaking waves are not merely water folding over itself. They are events that inject air into the ocean as a cloud of bubbles, creating an enormous temporary increase in surface area. Each bubble is a little moving interface with its own boundary layer. Bubbles rise, shrink, and sometimes dissolve; they can transport gases into the water far more efficiently than a flat surface could manage on its own. This is one reason the ocean's gas exchange does not scale gently with wind. Past a threshold where wave breaking becomes common, the effective exchange rate can jump. The sea, quite literally, froths with extra pathways.

The idea that wind controls exchange is now a staple of climate and ocean models, but the route to that idea ran through an unexpectedly dramatic mid-century marker: radiocarbon from nuclear weapons testing. In the 1950s and early 1960s, atmospheric tests produced a spike of radioactive carbon-14, sometimes called "bomb radiocarbon," that spread through the atmosphere and began to enter the ocean. This accidental tracer allowed oceanographers to estimate how quickly carbon dioxide and related carbon species were exchanged at the surface. It turned out that the ocean's uptake could not be explained without acknowledging a strong dependence on wind and the turbulence it generates. The planet had, in effect, conducted a grimly unplanned experiment on itself, and the oceanographic community learned to read the results in the pattern of radiocarbon in seawater.

Wind is not the only driver. Temperature quietly sets the stage by controlling *solubility*, the basic capacity of water to hold a gas. Cold water can hold more gas than warm water, just as a cold soda stays fizzy longer than a warm one. This is not a small effect. For oxygen, solubility changes strongly with temperature, which is one reason the cold high-latitude oceans are naturally oxygen-rich at the surface. For carbon dioxide, temperature matters too, though the chemistry is more involved because dissolved CO_2 participates in reactions that create bicarbonate and carbonate. Still, as a first intuition, warming pushes gases out and cooling pulls them in.

This temperature control has a seasonal rhythm that can be felt far beyond the laboratory. In many mid-latitude regions, surface waters cool in winter, increasing solubility and allowing the ocean to take up more oxygen and carbon dioxide if the air supplies it. In summer, warming reduces solubility and can encourage outgassing. But the story does not reduce to "warm means out, cold means in" because the boundary is also governed by mixing. Winter storms can deepen the surface mixed layer, stirring down gases absorbed at the surface and making room for more. Summer stratification can trap gases near the surface, allowing them to equilibrate more quickly with the atmosphere and limiting further uptake. Wind and temperature do not act independently; they negotiate through the structure of the surface ocean.

Turbulence is the third partner in this negotiation, and it is the one most likely to be underestimated by intuition because turbulence is messy. People often imagine a windy ocean as "mixed," but the relevant mixing for gas exchange is not only the stirring of the top few metres; it is the micro-turbulence near the surface that governs how thin the diffusive boundary layer becomes. A gentle breeze can roughen the surface without generating strong shear at the millimetre scale. A sudden gust can do the opposite: it can briefly intensify shear and produce a surge of exchange disproportionate to its duration. This sensitivity to intermittency is one reason why measuring gas fluxes at sea has been so challenging. A single afternoon of measurements can look unrepresentative, not because it is wrong, but because the ocean's breathing is punctuated.

A concrete example comes from oxygen on stormy nights. Oxygen is often used as a "test gas" for exchange processes because it is abundant, biologically important, and measured with high precision. During storms, especially in cold regions, breaking waves can inject bubbles and drive oxygen rapidly into surface waters that are undersaturated, sometimes following a period of biological consumption or upwelling of low-oxygen water. The ocean can switch from a gentle exchange regime to a forced-breathing regime in hours. Fishermen may not think in terms of transfer velocity, but they know the sea changes character after a blow: water can become clearer or murkier, plankton layers can be disrupted, and the surface can feel newly "alive" with foam and spray. Embedded in those changes is a sharp alteration in the rate at which oxygen and carbon dioxide are being moved.

The micrometre-scale boundary layer might seem too small to matter for a planet, yet it acts like a bottleneck in a global industrial system. The atmosphere contains on the order of 10^{19} moles of gas; the ocean contains vastly more water. Whether the ocean can absorb a significant fraction of anthropogenic carbon dioxide over years and decades depends on whether the surface can pass carbon-bearing molecules inward fast enough, and whether circulation can then carry that carbon away from the surface. Gas exchange is the first tollbooth on a long highway. If the tollbooth is slow, traffic backs up in the atmosphere.

The physical chemistry of this tollbooth begins with partial pressure. A gas in air exerts a partial pressure proportional to its fraction of the atmosphere. In water, gases have an equilibrium concentration related to partial pressure through Henry's law: for a given temperature (and salinity), there is a predictable concentration that corresponds to equilibrium with the air. If the actual dissolved concentration is lower than equilibrium, the ocean tends to absorb the gas; if higher, it tends to release it. This difference is often described by a Δp or ΔC: a gradient that provides the "desire" for flux. Transfer velocity provides the "ability."

One can get surprisingly far with the simple flux idea

$$F = k\,\Delta C,$$

where F is the flux (moles per square metre per time), k is transfer velocity (metres per time), and ΔC is the difference between the concentration implied by atmospheric conditions and the measured concentration in the surface water. The equation is almost too neat, and that neatness can be misleading. It encourages the belief that k is a stable property of the ocean, a number one can look up like the density of seawater. In reality, k is a summary of weather.

Even the act of deciding which side of the interface limits exchange depends on the gas. For sparingly soluble gases, the water side often dominates resistance; for more soluble gases, the air side can matter more. Carbon dioxide has an extra twist: it reacts with water and with hydroxide ions, meaning that chemical processes can effectively speed its uptake under certain conditions, a phenomenon called *chemical enhancement*. The surface is therefore not only a mechanical gate but also a reactive boundary.

Yet mechanics still hold the starring role because chemistry cannot proceed without contact. The ocean and atmosphere only "meet" over the brief lifetime of eddies and waves. A flat, glassy sea is a poor meeting place. A roughened surface is a crowded marketplace, with countless microscopic encounters.

Consider what happens during a day of rising wind. At first, capillary waves appear—small ripples held together by surface tension. These ripples increase surface area slightly and, more importantly, signal that wind stress is being transmitted to the water. Shear develops at the surface, and tiny turbulent motions begin to erode the diffusive layer. As wind builds, gravity waves grow; they travel and interact, and their crests sharpen. Whitecaps begin to appear where crests topple. Each breaking event injects bubbles and creates a turbulent wake that extends downward. Transfer velocity climbs. By the time the sea is covered in frequent whitecaps, air is being hammered into water and water is being flung into air. Exchange becomes limited less by the patience of diffusion and more by the intensity of the violence at the surface.

Bubbles deserve particular attention because they complicate the comforting symmetry of diffusion. One might imagine that bubbles simply accelerate equilibration: if the air has more oxygen than the water, bubbles help oxygen enter; if the water has more, bubbles help it leave. But bubbles can introduce asymmetry. As bubbles are forced downward by wave breaking, they experience higher pressure; gases compress and can dissolve more readily into the surrounding water. When those bubbles later rise, they may have already surrendered some of their gas load. In effect, bubble injection can create a net downward pump for certain gases under certain conditions, even beyond what one would expect from the simple surface gradient alone. This is not a separate process from gas exchange; it is gas exchange mediated by a particular geometry: curved interfaces moving through pressure gradients.

The ocean's breathing is therefore not just a surface story; it has a vertical reach, sometimes only centimetres, sometimes metres, depending on conditions. That reach matters because it determines whether absorbed gases linger at the surface long enough to be blown back out. A shallow, strongly stratified surface layer may quickly become saturated with oxygen or carbon dioxide, slowing further uptake. A deeper mixed layer can store more without saturating, maintaining a gradient that keeps flux going. Wind can deepen the mixed layer; cooling can deepen it too by making surface waters denser and prone to sink. Warming and freshening (from rain or ice melt) can sharpen stratification, thinning the mixed layer and weakening this storage effect. The breath of the waves is coupled to the ocean's habit of layering itself.

A historical anecdote captures both the ingenuity and the frustration of trying to measure these processes directly. Early attempts at quantifying air–sea fluxes sometimes used floating chambers: small domes placed on the sea surface to trap air and measure changes in gas concentration over time. The chambers were conceptually straightforward, but the ocean is rarely cooperative. Chambers can alter the turbulence they are meant to observe, shielding the surface from wind and changing the very exchange rate under study. Later methods used tracers such as sulfur hexafluoride, deliberately released in patches, to track gas transfer without a physical barrier. The tracer patch would spread and dilute while its concentration declined partly through exchange with the air; by following that decline, researchers could infer transfer velocities. These experiments revealed again and again that wind speed alone is not enough; wave state and short-term variability matter.

The everyday relevance of all this can be easy to miss because the gases involved are invisible. But they make their presence known in indirect ways. When coastal upwelling brings cold, carbon-rich water to the surface, the ocean can become a source of CO_2 to the atmosphere even as it remains a net sink globally. The upwelled water may be supersaturated in CO_2 relative to the air; strong winds that drive upwelling also enhance transfer velocity, allowing rapid outgassing. The same physics that helps the ocean absorb anthropogenic carbon in some places can help it release natural carbon in others. The "lungs" inhale and exhale unevenly.

Or consider oxygen again, this time from the point of view of living things. A coastal embayment can become oxygen-poor in summer when stratification traps bottom waters and biological respiration consumes oxygen. A storm that breaks the stratification and drives air–sea exchange can rapidly re-oxygenate the surface and sometimes the whole water column, offering temporary relief to marine life. Yet if the storm also stirs up nutrients and fuels a bloom that later decays, it can worsen oxygen depletion in the longer term. Physical exchange is not a moral force; it is a set of levers that biology can pull and be pulled by.

One reason the ocean's gas exchange matters for climate is that it determines how quickly the ocean can respond to atmospheric changes. When atmospheric CO_2 rises, the ocean does not instantly adjust. The surface begins to absorb more, but the rate depends on transfer velocity, and the ability to keep absorbing depends on moving carbon away from the surface into the interior by circulation and mixing. A calm, strongly stratified ocean surface can become saturated relative to the atmosphere in a thin layer, throttling uptake. A windy, well-mixed surface can keep drawing down more. The weather, in other words, has a say in the global carbon budget.

At the scale of days, this can feel like a detail. At the scale of decades, it becomes an influence on how much warming occurs for a given set of emissions. The ocean does not merely store heat; it also stores carbon, and the gatekeeping begins with the mechanics of the surface.

The transfer velocity idea also helps explain why the ocean's exchange is patchy. The Southern Ocean, for example, is famously windy. It is also cold, which increases solubility, and it experiences strong mixing and deep winter overturning in places. One might therefore expect it to be a powerful sink for atmospheric CO_2, and indeed it is important. But it is also a region where deep waters, rich in dissolved inorganic carbon accumulated over long times, are brought to the surface. That upwelling can oppose the sink by supplying carbon that wants to escape. The balance between these tendencies is delicate, and it is set by the coupling of wind-driven physics and carbon chemistry. Even without diving into the chemistry, the physical lesson is clear: a windy ocean is not automatically an inhaling ocean. Wind is an amplifier. It amplifies whichever direction the gradient points.

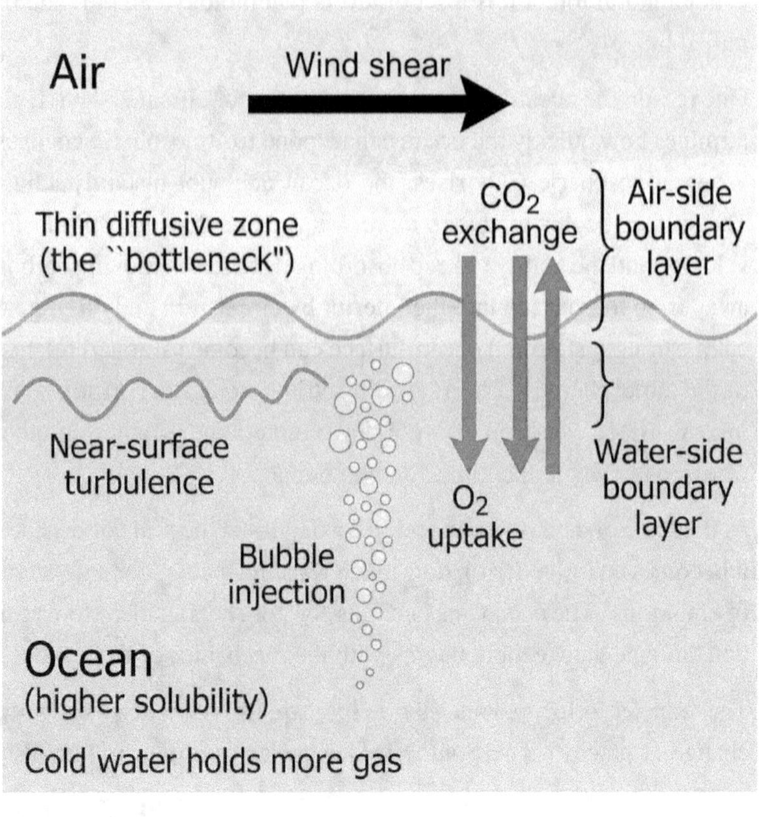

To bring these abstractions down to a single square metre of sea surface, imagine a patch of ocean under a passing weather front. Ahead of the front, the wind is light; the surface warms slightly under thin sunlight; the mixed layer is shallow. Carbon dioxide exchange is modest. As the front arrives, wind increases quickly. Whitecaps appear. The surface cools under rain and heat loss. The mixed layer deepens as the ocean is stirred and cooled, pulling surface water down and bringing water from below up to the surface. Suddenly, there is more capacity near the surface for gases to dissolve without saturating. Transfer velocity rises. If atmospheric CO_2 is higher than the water's equilibrium concentration, uptake accelerates. If the surface water was already rich in CO_2 because of upwelling or biology, outgassing accelerates instead. The same storm can produce opposite CO_2 fluxes in two nearby regions simply because the pre-existing gradients differ.

There is a final subtlety worth keeping in mind: the ocean surface is not chemically uniform, even when it looks uniform. The top millimetres can develop a *cool skin* or a *warm layer* depending on heat fluxes, and surfactants from biology can modify surface tension and dampen small-scale waves, changing turbulence and gas transfer. A slick of organic material can smooth capillary waves and suppress exchange locally, while nearby, a patch of breaking waves can enhance exchange dramatically. The planet's lungs are not an even fabric; they are stitched from patches that widen and narrow with weather and life.

The ocean, then, is not breathing in the slow, gentle way a lake might. It is breathing through a surface that is constantly being rebuilt by wind and waves, with temperature setting how much gas the water can hold and turbulence deciding how quickly the surface can be renewed. The air–sea boundary is a place where the large and the small meet: global cycles hinging on micrometres, climate-relevant fluxes hinging on a gust. When the wind rises and the sea whitens, the interface stops behaving like a boundary and starts behaving like a machine—one powered by weather, turning the atmosphere and ocean into a coupled system that can neither fully separate nor fully merge, only exchange.

6.2 The Dissolving Shield

The ocean has been doing humanity a quiet favour. Each year, a large share of the carbon dioxide released by burning coal, oil, and gas does not remain in the air. Some of it is taken up by land plants and soils, and a substantial fraction enters seawater. That uptake slows the rise of atmospheric CO_2 and, with it, the pace of warming. From a distance it can look like a simple act of dilution: add CO_2 to air, some dissolves into the sea, problem softened.

Up close, it is a chemical chain reaction with consequences. The ocean does not merely *hold* added carbon dioxide; it *transforms* it. Those transformations alter seawater's acidity, its inventory of carbonate ions, and the ease with which organisms can build shells and skeletons. The same chemistry that makes the ocean an effective absorber also makes it vulnerable to becoming less hospitable.

A good starting point is the familiar fizz in a carbonated drink. When you open a bottle, the pressure drops, dissolved CO_2 escapes, and bubbles form. In the ocean, the "bottle" is the atmosphere. As the concentration of CO_2 in air rises, the ocean's surface waters are pushed toward holding more dissolved CO_2. Unlike a drink, however, seawater is not just water. It contains dissolved salts and, crucially, a set of carbon-bearing ions that can rearrange themselves. The ocean's carbon is not stored as a single substance but distributed among several chemical forms. That distribution is the heart of the dissolving shield.

Chemists often group the dissolved carbon forms under one umbrella: *dissolved inorganic carbon*, abbreviated DIC. DIC includes aqueous carbon dioxide $CO_2(aq)$ (including a closely related form, carbonic acid), bicarbonate HCO_3^-, and carbonate CO_3^{2-}. In today's seawater, most DIC is bicarbonate. Carbonate is a smaller fraction, but it plays an outsized role because it is the form that combines with calcium to make calcium carbonate minerals: the substance of coral skeletons, many planktonic shells, and countless grains of marine sand.

The transformations linking these forms are well known, but they are easy to misunderstand because they run in both directions and depend on conditions. When CO_2 enters the ocean, some of it remains as dissolved $CO_2(aq)$, but a significant portion reacts with water:

$$CO_2 + H_2O \rightleftharpoons H_2CO_3$$

Carbonic acid H_2CO_3 is a fleeting participant, quickly dissociating:

$$H_2CO_3 \rightleftharpoons H^+ + HCO_3^-$$

and bicarbonate can dissociate further:

$$HCO_3^- \rightleftharpoons H^+ + CO_3^{2-}.$$

These reactions introduce a key character: the hydrogen ion H^+. The concentration of H^+ is what pH measures. pH is defined as

$$\mathrm{pH} = -\log_{10}[H^+]$$

so a small change in pH corresponds to a percentage change in hydrogen ion concentration. The numbers can seem deceptively modest: a shift from 8.2 to 8.1 looks tiny. On the logarithmic pH scale, it is not. It corresponds to a noticeable increase in $[H^+]$, meaning the water is measurably more acidic.

Here is the central twist. Adding CO_2 to seawater increases $[H^+]$, which lowers pH. But at the same time, those added hydrogen ions tend to combine with carbonate ions:

$$H^+ + CO_3^{2-} \rightleftharpoons HCO_3^- .$$

So as $[H^+]$ rises, $[CO_3^{2-}]$ falls, and bicarbonate increases. The ocean takes in carbon partly by shifting its carbon inventory into bicarbonate. That shift is one of the reasons the ocean can absorb such a large amount of CO_2 without immediately becoming saturated with dissolved CO_2. It is also why uptake carries a price: the carbonate ion pool is drawn down, and carbonate is the raw material for calcification.

The term *buffer* is often used to describe systems that resist changes in pH. Seawater is buffered: despite absorbing vast quantities of CO_2, its pH remains well above neutral (7). The buffering comes largely from the carbonate system itself and from the ocean's *alkalinity*, a measure of the seawater's capacity to neutralize acids. Alkalinity is not a single substance; it is a bookkeeping of charges, dominated in seawater by bicarbonate and carbonate. In everyday terms, alkalinity is the ocean's chemical "stiffness" against swings in acidity.

But buffering is not a force field. It does not block change; it shapes it. When CO_2 is added, pH declines rather than crashing, yet the decline is real, and the accompanying loss of carbonate ions is often more biologically consequential than the pH number alone. A shell-building organism does not experience pH as an abstract scale. It experiences the availability of carbonate ions and the tendency of calcium carbonate to precipitate or dissolve in its immediate environment.

That tendency is often expressed with a quantity called the *saturation state*, written Ω, for a particular mineral form such as aragonite or calcite (two crystalline forms of calcium carbonate). In simplified form,

$$\Omega = \frac{[\mathrm{Ca}^{2+}][\mathrm{CO}_3^{2-}]}{K_{sp}},$$

where K_{sp} is a solubility product that depends on temperature, salinity, and pressure, and where calcium is relatively abundant and fairly stable in seawater. Carbonate concentration, by contrast, is the variable that shifts strongly with added CO_2. When $\Omega > 1$, seawater is supersaturated and precipitation is thermodynamically favoured; when $\Omega < 1$, it is undersaturated and dissolution is favoured. Many organisms can calcify even when conditions are not ideal because biology can manipulate local chemistry, but the energetic cost tends to rise as Ω declines.

This is where the "shield" metaphor earns its second half. By absorbing CO_2, the ocean partly shields the atmosphere. Yet the chemical rearrangement that enables uptake corrodes a different kind of protection: the ocean's capacity to keep calcium carbonate stable and plentiful. The shield dissolves from within.

A historical note helps show how this became visible as more than a laboratory curiosity. In the late twentieth century, as high-precision measurements of carbon chemistry became routine, oceanographers began to establish time-series stations in the open ocean. One of the most famous is near Hawai'i: Station ALOHA, associated with the Hawaii Ocean Time-series. Over decades, careful measurements there have shown a steady increase in dissolved inorganic carbon and a corresponding decline in pH in surface waters, tracking the rise of atmospheric CO_2 with remarkable fidelity. The significance is not that the water became "acidic" in the colloquial sense; it remains alkaline. The significance is that a long-lived, well-buffered system is nonetheless being pushed in one direction consistently enough that the trend rises above natural variability.

The same story appears in other basins and in independent observing programs. The Intergovernmental Panel on Climate Change has assessed that surface ocean pH has declined since preindustrial times largely because of the ocean's uptake of anthropogenic CO_2, with observed declines at long time-series sites on the order of a few hundredths of a pH unit per decade since the late 1980s. The numbers are small per year and large per human lifetime.

The chemistry also reveals why the ocean cannot simply keep absorbing CO_2 at the same relative pace forever, at least not without consequences. There is a measure of the ocean's resistance to absorbing more CO_2 called the *Revelle factor*, sometimes called the buffer factor. It describes how strongly the partial pressure of CO_2 in seawater changes in response to a change in dissolved inorganic carbon, for a given alkalinity. A higher Revelle factor means that a small addition of DIC produces a relatively large increase in seawater pCO_2, bringing the surface closer to equilibrium with the atmosphere and reducing the drive for further uptake. In plainer terms: as the ocean takes up carbon, it becomes less willing to take up additional carbon, and the chemical shift that enables uptake also reduces future capacity.

Temperature threads through this in a way that can be counterintuitive. Warmer water holds less CO_2 physically, which tends to reduce uptake. But temperature also changes the balance among carbon species and alters the solubility products that define saturation states. Cold waters can hold more dissolved CO_2, which might sound beneficial for uptake, yet cold high-latitude waters are also where carbonate saturation states can be naturally lower, making those regions especially sensitive to further declines in $[CO_3^{2-}]$. Pressure matters too: calcium carbonate becomes more soluble at depth, which is why there is a depth horizon below which certain minerals tend to dissolve. As anthropogenic carbon penetrates into the ocean interior, these dissolution horizons can move upward, a process often called *shoaling*. A horizon that once sat safely deep can rise toward the habitats of organisms living on continental shelves or in the upper ocean.

The living consequences become especially vivid in upwelling systems, where physics and chemistry cooperate in a less comforting way. Along parts of the west coast of North America, winds can drive surface waters offshore, allowing deeper waters to rise and replace them. Those deeper waters are often rich in dissolved inorganic carbon and low in oxygen because they have accumulated the products of respiration: organic matter sinking from the surface is consumed by microbes, producing CO_2 and using oxygen. When such waters reach the surface, they can be naturally high in CO_2 and lower in pH than surface waters elsewhere, even before adding the modern anthropogenic burden. In these regions, the "baseline" is already close to thresholds that matter for calcifiers.

A concrete example is the experience of shellfish hatcheries in the Pacific Northwest of the United States in the 2000s, when episodes of low-pH, high-CO_2 seawater were linked to failures in oyster larvae production. The larvae were trying to build shells at a life stage when their margin for chemical difficulty is thin. The oceanographers and hatchery operators who worked together on the problem did not need metaphors: they needed real-time measurements of carbonate chemistry and operational strategies, such as timing water intake to avoid corrosive pulses or modifying water chemistry in controlled settings. The broader significance is that "ocean acidification" is not a distant, uniform trend; it can arrive as intermittent shocks in particular places, layered on top of a long-term drift.

The word *acidification* itself can mislead. People sometimes imagine the ocean turning into vinegar. Seawater is still alkaline, and will remain so under plausible scenarios. Acidification means movement toward lower pH, not crossing the neutral point. A more revealing phrase might be "carbonate depletion," because it captures the mechanism that links atmospheric CO_2 to biological building materials. Yet pH remains a powerful public symbol because it is a single number, easy to chart and easy to compare. The chemistry insists on holding both ideas at once: a modest pH shift can represent a large change in hydrogen ion concentration, and a large ecological effect can occur without the water becoming acidic in the everyday sense.

The carbonate system can also be understood as a set of choices forced by added CO_2. In a simplified view, there is a pool of DIC that can be distributed among $CO_2(aq)$, HCO_3^-, and CO_3^{2-}. When more CO_2 is added, the system shifts toward forms that "store" the carbon while maintaining electrical balance. Bicarbonate is a stable compromise, so it grows. The price is paid by carbonate, which shrinks, and by pH, which falls. The ocean's buffering is therefore not the elimination of consequences; it is the channeling of consequences into particular chemical directions.

Placed alongside the earlier physical story of air—sea exchange, this chemistry also clarifies why speed matters. When wind and turbulence accelerate the transfer of CO_2 across the surface, they do not simply increase the amount of dissolved CO_2 in water. They feed carbon into a reactive network. That network then determines how much stays as $CO_2(aq)$, how much becomes bicarbonate, how much carbonate is consumed, and how pH responds. The surface ocean can therefore absorb CO_2 rapidly during windy, cold periods, but the chemical consequences of that absorption persist long after the storm has passed.

The same is true in reverse. If a parcel of water rich in dissolved CO_2 reaches the surface, strong winds can accelerate outgassing, but the carbonate system again shapes the response. As CO_2 leaves the water, $[H^+]$ declines, pH rises, and carbonate ions can recover. The system is reversible in principle, but it is being driven persistently in one direction by the sustained rise of atmospheric CO_2. The ocean is not simply oscillating around a stable state; it is being marched.

The chemical pathway can be sketched without drowning in detail, and doing so makes the logic hard to ignore:

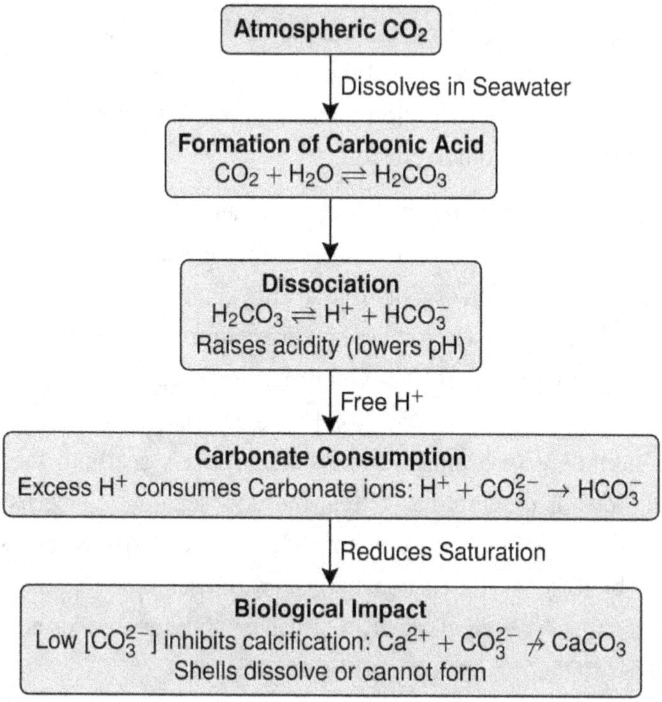

In this minimalist diagram, the moral is not that carbon dioxide "turns into acid" in a simplistic way, but that every added molecule of CO_2 increases the tendency to produce hydrogen ions and to convert carbonate into bicarbonate. The ocean's chemistry is being steered toward a state with less carbonate available for building and maintaining calcium carbonate structures.

The dissolving shield also has a deeper timescale that often goes unmentioned. Over thousands of years, the ocean can regain some of its buffering capacity through processes that add alkalinity, including the weathering of rocks on land and the dissolution of carbonate sediments on the seafloor. In a geological sense, Earth has a thermostat-like mechanism: added CO_2 can eventually be balanced by rock weathering and burial of carbon. The problem is speed. Human emissions have risen within a couple of centuries, and the ocean's fast response is absorption and chemical rearrangement in the surface and upper layers, not a rapid replenishment of alkalinity. The quick shield comes with quick erosion.

There is also a human dimension that chemistry brings into focus. Much of the planet's protein comes from the sea, and many coastal economies depend on organisms that either calcify themselves or rely on calcifying ecosystems. Coral reefs are tourist destinations, storm buffers, and nurseries for fish. Pteropods—tiny plankton sometimes called "sea butterflies"—are a food source for larger animals in some high-latitude ecosystems. Oysters and mussels are both food and livelihood. It is tempting to treat these as separate concerns from atmospheric carbon, as if the atmosphere were "climate" and shells were "biology." The carbonate system refuses that separation. It makes CO_2 a chemical agent in seawater, linking smokestacks and tailpipes to the microscopic arithmetic of ions.

If the ocean's surface is the gate through which carbon enters, the carbonate system is the accounting office that decides where that carbon sits and what it does to the balance of acidity and carbonate availability. That accounting is not neutral. It turns the ocean's generosity in absorbing CO_2 into a cost paid in altered chemistry, felt most sharply where the ocean is already cold, already upwelled, or already close to the edge of calcium carbonate stability. The shield still works, but it thins, molecule by molecule, in water that remains blue and familiar even as its internal chemistry is rewritten.

6.3 Suffocating Depths

Oxygen feels like the simplest gift the ocean receives from the atmosphere: invisible, plentiful, and essential. At the surface it is renewed constantly, pushed across the air—sea interface by wind and wave-driven exchange. The puzzle begins a few tens to a few hundreds of metres down, where sunlight fades, photosynthesis wanes, and the ocean's interior starts to look less like a well-ventilated room and more like a crowded hall with too few doors.

In many regions there is a layer where oxygen concentrations drop to remarkably low values compared with the surface. Oceanographers call these regions *Oxygen Minimum Zones*, or OMZs. They are not empty voids; they are habitats with their own adapted communities, microbial pathways, and boundaries that animals learn to avoid. But as the climate warms, these low-oxygen regions are changing in ways that matter to fisheries, ecosystems, and the ocean's role in regulating greenhouse gases. "Deoxygenation" can sound like an abstract, global statistic. In the water, it can mean a habitat shrinking into a thinner band, a migration corridor pinching shut, or a night-time ascent for prey that predators can no longer follow.

A first misunderstanding to clear away is the idea that low oxygen must mean "pollution" or local stagnation. Coastal dead zones caused by nutrient runoff are real and serious, but the open-ocean OMZs that span thousands of kilometres are created by the ocean's ordinary workings. They appear most strongly in the eastern tropical Pacific, the northern Indian Ocean, and parts of the tropical Atlantic—places where circulation patterns, biology, and physics conspire to remove oxygen faster than it can be replaced.

The conspiracy begins at the surface. Sunlit waters grow phytoplankton, the microscopic plants that convert CO_2 into organic matter. Some fraction of that organic matter sinks as crumbs, clumps, and fecal pellets. This raining carbon is the ocean's biological carbon pump, a process that helps store carbon away from the atmosphere. Yet carbon sinking is also oxygen sinking, in a sense: when microbes and other organisms consume that organic matter in the dim interior, they use oxygen to "burn" it back into CO_2. The ocean interior is a vast respiration chamber. Even without any human influence, this respiration creates midwater layers where oxygen is depleted.

Whether that depletion becomes extreme depends on ventilation: the rate at which oxygenated water from the surface is transported downward and sideways. The deep ocean is not ventilated everywhere at once. It is renewed in particular regions where surface waters become dense and sink, carrying oxygen with them. Elsewhere, large volumes of water can remain isolated from the surface for decades to centuries. In the meantime, respiration quietly draws down oxygen.

The OMZ often sits at the intersection of two patterns: a productive surface that supplies plenty of sinking food for microbes, and a sluggish interior circulation that fails to replenish oxygen quickly. The upper boundary is shaped by mixing and by the depth of the surface mixed layer; the lower boundary by the slow arrival of oxygen from deeper currents. In between lies a "shadow zone" where microbes do most of the work and larger animals must ration their choices.

Warming alters this system through two physical levers and one biological one. The first lever is solubility. Warm water holds less oxygen than cold water, just as warm soda goes flat. That effect is immediate and unavoidable: for the same atmospheric oxygen, a warmer surface ocean equilibrates to a lower dissolved concentration. This does not by itself create a minimum zone, but it lowers the baseline from which the interior draws.

The second lever is stratification, the ocean's tendency to arrange itself in layers that resist mixing. When the surface warms, it becomes lighter relative to deeper waters. A lighter surface floats more stably, and stable layering suppresses the vertical exchange that would otherwise deliver oxygen downward. Freshwater inputs can strengthen stratification too. Stronger stratification is like closing the vents in a building: oxygen enters at the roof, but less of it makes it into the hallways below.

The third lever is respiration. Microbial consumption of organic matter depends on temperature and on the supply of food. In general, metabolic rates rise with temperature, so warming can encourage faster oxygen consumption, provided organic matter is available. At the same time, climate-driven changes in nutrients and productivity can alter how much organic matter sinks. Different regions can therefore respond differently: in some places the biological "demand" for oxygen may rise, in others the supply of sinking organic carbon may shift. But the broad tendency is troublingly coherent: warmer, more stratified waters are less capable of delivering oxygen into the interior, while respiration continues to draw oxygen down.

These influences do not simply lower oxygen everywhere uniformly. They reshape the geometry of oxygen in the water column, often expanding low-oxygen volumes and raising the upper boundary of the OMZ toward the surface. That upward movement matters more than it might sound. The upper few hundred metres of the ocean contain much of its living activity: the daily feeding migrations of small fish and squid, the hunting routes of tuna, the growth zones of larvae. If the oxygen-poor layer creeps upward, it can compress the habitable region into a thinner band between "too little oxygen" below and "too warm or too nutrient-poor" above. Oceanographers sometimes call this *habitat squeeze*. For animals that must keep moving to find food, a squeeze is not a mild inconvenience; it can be a rewrite of the map.

A concrete example often discussed by marine ecologists involves tuna and billfish. These are high-metabolism predators that require well-oxygenated water to sustain fast swimming. In parts of the eastern tropical Pacific, oxygen-poor waters already lie relatively shallow. As the upper boundary of the OMZ shoals, the depth range available to such predators can narrow. Their prey, sometimes more tolerant of low oxygen, may still occupy deeper layers. Predators can become constrained to the surface where temperature stress may be higher or where fishing gear is more concentrated. Fisheries are then reshaped not only by changes in abundance but by changes in vertical distribution—a three-dimensional problem that turns into economics.

OMZs also matter for chemistry beyond oxygen itself. When oxygen becomes scarce, microbes switch to other electron acceptors to continue extracting energy from organic matter. Nitrate becomes especially important, leading to processes such as denitrification and anammox that remove fixed nitrogen from the ocean. Fixed nitrogen is a limiting nutrient for productivity in many regions; losing it can alter the biological engine that supports food webs. These low-oxygen pathways can also produce nitrous oxide N_2O, a potent greenhouse gas. An expanding OMZ can therefore change the ocean's role not only as a sink or source of CO_2 but as a participant in other climate-relevant cycles.

The story has a revealing historical thread: the ability to map oxygen in the ocean interior improved dramatically with the rise of systematic ocean surveys and, later, autonomous floats. Early expeditions measured oxygen with painstaking chemical titrations, and the resulting maps showed the broad outline of OMZs long before climate change entered the conversation. Those early measurements were enough to establish that large, persistent low-oxygen layers are normal features of the ocean. What has changed is the precision and continuity with which oxygen can be monitored. Modern sensors on research ships and on profiling floats now record oxygen across seasons and over wide areas. This increase in observational power has made it harder to dismiss deoxygenation as mere anecdote, even while it has also revealed something that complicates the narrative: natural variability can be large.

In some regions, oxygen levels swing with multidecadal rhythms tied to changes in winds, currents, and the great climate modes of the ocean-atmosphere system. A period of stronger ventilation can temporarily raise oxygen in the interior; a period of weaker ventilation can lower it. This variability matters for two reasons. First, it means a short time series can be misleading: a decade of decline might be followed by partial recovery, not because the underlying warming influence has stopped, but because the ocean's internal rhythms have turned. Second, it means attribution—the effort to separate human-driven trends from natural cycles—is challenging in particular basins, especially where observations have been sparse historically. Scientists have learned to be cautious: the ocean can hide long-term change inside noisy regional swings.

Caution, however, is not comfort. Multiple lines of evidence indicate that the global ocean has been losing oxygen over recent decades and that low-oxygen zones are expanding in several locations. The Intergovernmental Panel on Climate Change has assessed observed oxygen loss and projects further declines under continued warming, with expansions of OMZ volume in higher-emissions futures. What makes this especially sobering is the timescale. Oxygen loss linked to warming and circulation can persist for a long time. Once the interior ocean has been altered, it cannot be rapidly "re-oxygenated" simply by a few years of calmer weather. Ventilation is slow. The deep ocean remembers.

It helps to picture oxygen not as a static property but as a budget.

Oxygen enters the ocean primarily at the surface, then is transported into the interior by mixing and currents. Oxygen is consumed in the interior by respiration and by chemical reactions involving reduced substances. Where supply is high and consumption low, oxygen remains plentiful. Where consumption outruns supply, oxygen declines. OMZs are the places where the budget is most unbalanced.

The surface, by contrast, can give a false sense of security. The top layer often remains reasonably oxygenated, continually recharged by the atmosphere. Someone on a ship can sample surface water and find oxygen near saturation while, a few hundred metres down, oxygen has fallen to a fraction of that value. The ocean can look healthy at the skin while suffocation spreads below.

A second false intuition is that oxygen should simply decline steadily with depth. In many regions it does not. Oxygen can be high at the surface, then drop rapidly beneath the mixed layer to an OMZ, and then rise again in deeper waters that were last ventilated in cold, high-latitude regions and have retained some oxygen despite age. This non-monotonic profile is one reason OMZ boundaries can be sharp and ecologically meaningful. Animals can sometimes find a “ledge” of acceptable oxygen at depth, then encounter a sudden wall where it becomes too low.

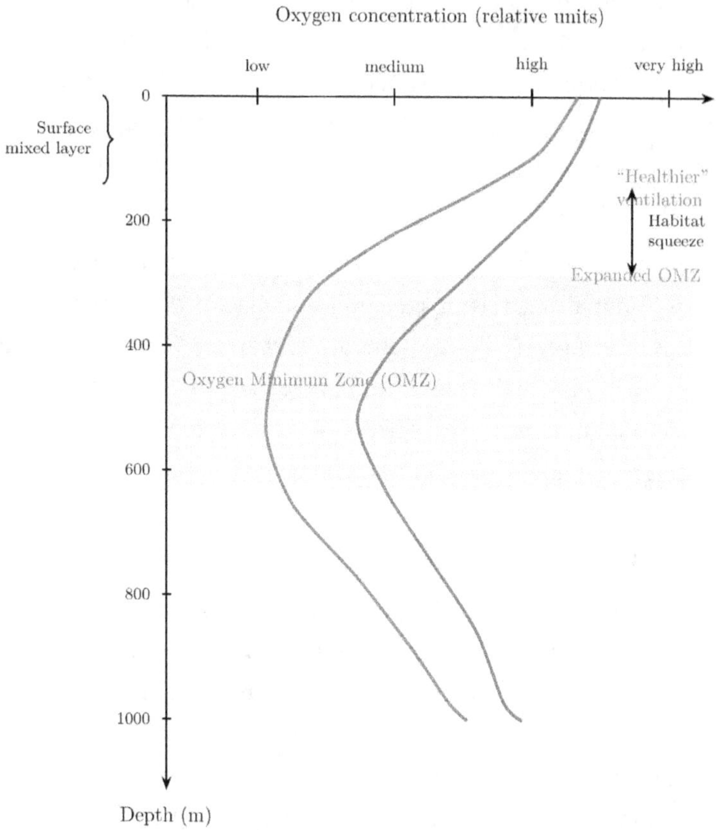

This schematic is not a specific dataset; it is a way of seeing the problem. The "expanded OMZ" profile shows a shallower onset of low oxygen and a deeper recovery, broadening the low-oxygen band. Even if surface oxygen remains relatively high, the usable midwater space tightens.

The biological consequences ripple outward. Low oxygen can slow growth, alter reproduction, and change behaviour. Some animals become less able to escape predators or to capture prey. Others gain an advantage if they can tolerate hypoxia. Jellyfish, for example, often cope better with low oxygen than many fish, and there is concern in some regions that deoxygenation could tilt ecosystems toward gelatinous dominance, with knock-on effects for fisheries and carbon cycling. Microbial communities also reorganize, favouring pathways that thrive in oxygen-poor conditions and reshaping nutrient availability.

There is also an uncomfortable coupling with warming itself. Many marine animals face rising metabolic demand in warmer water: their bodies "run faster." At the same time, warmer water contains less oxygen and may deliver less from the surface to depth. That double pressure is sometimes described as a narrowing of the "oxygen and capacity" window—the gap between oxygen supply and oxygen demand. Even without invoking detailed physiology, the intuition is direct: higher temperatures ask organisms to do more with less.

A striking feature of OMZ expansion is that it can be both gradual and sudden. Gradual, because long-term warming and stratification trends slowly alter ventilation. Sudden, because the boundary of an OMZ can move with seasons, with climate oscillations, or with episodic events such as strong upwelling years. For coastal communities and fisheries, the sudden manifestations can be the most memorable: a year when catches change, when fish are found unusually shallow, when hypoxic waters intrude onto shelves and cause die-offs. Those events are not always caused by global deoxygenation, but global deoxygenation can raise the odds and intensify the baseline, making extremes more consequential.

The ocean's oxygen story is sometimes told as a tragedy of the deep. Yet it is also a lesson in how tightly coupled the ocean is to the atmosphere and to the living carbon cycle. Oxygen is produced in the sunlit surface by photosynthesis, consumed in the dark by respiration, and redistributed by currents that respond to winds and heat. Warming alters solubility at the surface and alters stratification that governs ventilation. Meanwhile, the biological pump that exports carbon—so central to the ocean's role in buffering atmospheric CO_2—also exports the fuel for oxygen consumption at depth. The planet's chemistry does not give free gifts: the same machinery that helps regulate climate creates vulnerabilities when pushed.

If you want a single image to keep, it is not a lifeless "dead zone" spreading like ink. It is a boundary, sometimes sharp, that slides upward and outward. Above it, animals can sprint, hunt, and migrate with familiar freedom. Below it, they must slow down, choose different depths, or turn away entirely. The boundary is drawn not with lines on a map but with the balance of oxygen molecules: how many can be delivered, how many are consumed, and how quickly the ocean can replace what it spends.

In the dim interior, oxygen is a currency. The ocean has begun to spend it faster, while restocking becomes harder. The bill is paid in altered habitats, changed food webs, and a chemistry that quietly shifts toward states that would once have been confined to rarer corners of the sea.

Chapter 7

The Ocean's Memory

While the atmosphere forgets yesterday's weather in a week, the ocean retains thermal fingerprints for centuries, effectively remembering the climate of the Middle Ages. This chapter descends into the physics of the abyss to reveal how freezing poles drive a sluggish, massive global current that buffers the Earth against rapid temperature shifts.

7.1 The Engine in the Ice

In polar winter, the ocean does something that feels, at first glance, like a simple phase change: liquid water becomes solid water. Yet the moment seawater freezes, it reveals a quiet trick of chemistry. Ice is fussy. It prefers a crystal lattice of almost H_2O, and it makes little room for salt. So when sea ice forms, most of the dissolved salts are pushed out of the growing crystals and left behind in the liquid. Oceanographers call this *brine rejection.* The phrase sounds like a minor side effect. In fact it is one of the planet's most efficient ways of manufacturing density, and density is the currency that buys motion in the deep ocean.

Seawater density depends mainly on temperature and salinity. Cold water is denser than warm water; salty water is denser than fresh water. At the surface, winds can push water around quickly, but winds cannot easily force the abyssal ocean to move. Deep currents are powered instead by tiny, persistent differences in density that act like a slow gravitational engine: heavy water sinks, lighter water rises, and the rest of the ocean must adjust around that vertical exchange. If you want to set a vast volume of water in motion without touching it directly, you change its density and let gravity do the work.

Sea ice formation is a remarkably concentrated way to do that, because it can raise salinity right where the ocean is already cold. Cooling alone can make surface water dense enough to sink, but there is a limit: once the surface water reaches the freezing point, it cannot cool further without turning into ice. Salting the water, by contrast, can continue as long as ice is forming. The freezing ocean therefore has a second lever to pull after temperature has bottomed out: it can keep increasing density by increasing salinity.

The physics becomes easiest to picture at the moment a new skin of ice starts to grow. The topmost ocean is close to the freezing point, often a little below 0°C because salt lowers water's freezing temperature. As ice crystals knit together, they trap only a small fraction of the salt. The rest is squeezed into the remaining liquid, turning it into brine: colder, saltier, denser than the water beneath. Dense fluids do what dense fluids always do in a gravitational field. They sink.

That sinking rarely happens as a neat sheet sliding downward. The ocean is turbulent, full of swirls and filaments. Brine released from ice tends to descend in plumes and fingers, mixing as it goes, but still carrying a density signature that can survive long enough to matter. Multiply that by days, then months, then years across broad polar shelves and ice edges, and the bookkeeping becomes enormous: a seasonal process near the surface can ventilate great volumes of the deep sea.

Cross-section: sea ice formation and sinking of dense brine-enriched water

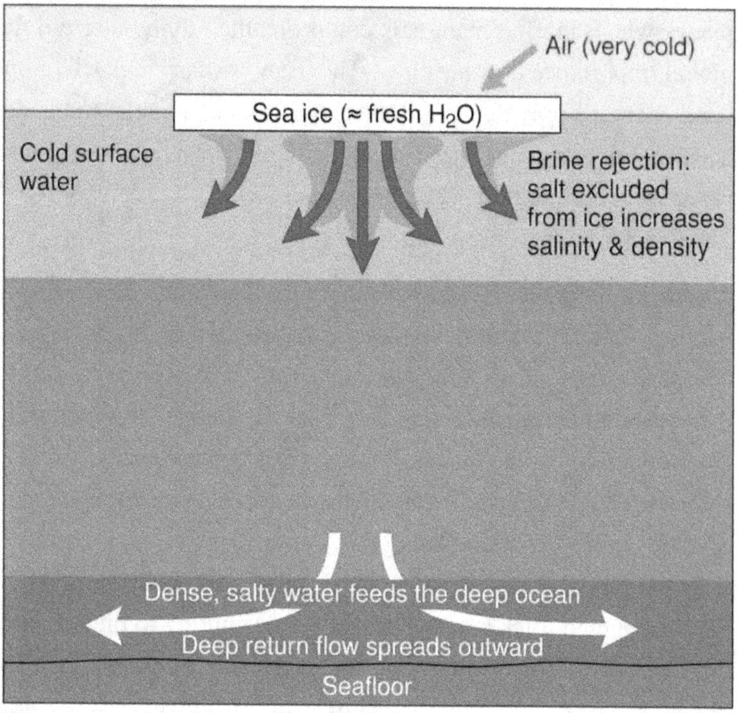

Brine rejection is not a curiosity confined to textbooks. It is written into the geography of polar seas. Consider the places where sea ice forms most vigorously: coastal polynyas, those oddly persistent openings in the winter ice that look, from above, like dark wounds in a white sheet. A polynya can be created by winds that blow newly formed ice away from the coast as fast as it appears, or by currents and tides that keep water moving and the surface from sealing over. Either way, the key feature is the same: open water keeps being exposed to bitter air, so it keeps freezing. Freezing keeps rejecting brine. Brine keeps making dense water.

For much of the twentieth century, polynyas were known to explorers as practical features. They could be hazards, offering a sudden stretch of open water in a landscape of ice; they could also be lifelines, places where marine mammals could breathe. Only later did their global importance become clear. The irony is sharp: a patch of open water, perhaps only tens of kilometers across, can matter to the deep ocean more than a far larger area of thick, unbroken ice. The open water is an efficient ice factory. The ice factory is a density factory. The density factory is a pump.

A concrete example sits along the margins of Antarctica, where frigid winds pour off the continent. When katabatic winds sweep down the ice slopes and out over the sea, they can drive sea ice away from the coast and maintain coastal polynyas. In these small regions, sea ice production can be intense enough to create very salty, very cold shelf waters. Those shelf waters may become dense enough to spill off the continental shelf, sliding down submarine canyons and slopes as gravity currents. As they descend, they entrain surrounding water, mixing and thickening, eventually contributing to the formation of Antarctic Bottom Water, one of the densest water masses in the global ocean. It is difficult to imagine a more dramatic scaling: local winds and microscopic crystal growth setting up currents that fill ocean basins.

The process is not always steady. Brine rejection can happen in bursts, and its effects are shaped by the local architecture of the ocean. Continental shelves matter because they provide a shallow staging ground where brine can accumulate and where cooling can act on a relatively small volume of water before that water is released downslope. The shape of the seabed matters because it guides the dense flows like rails. Sea ice matters not only as a salt filter but also as a lid: it changes how the ocean loses heat to the atmosphere and how winds stir the surface. Even ice shelves, floating extensions of the Antarctic ice sheet, can influence how water masses form by altering circulation and meltwater distribution. The "engine" in the ice is therefore not a single piston but a coupled machine of atmosphere, ice, and ocean, with brine rejection as a particularly powerful gear.

A historical anecdote helps capture how surprising these polar mechanisms once were. When the *Fram* drifted across the Arctic Ocean in the 1890s, Fridtjof Nansen's expedition aimed to use sea ice as a vehicle, letting it carry the ship with the slow drift of the pack. Nansen and his colleagues were careful observers, and among the most important things they noticed was not a heroic detail about navigation but a physical one: the ice drifted at an angle to the wind, not directly downwind. That observation helped inspire the later work of Vagn Walfrid Ekman, who explained how Earth's rotation shapes the response of the ocean to wind, creating a spiral of currents with depth. The anecdote matters here because it shows the broader pattern of discovery: polar ice, once treated mainly as an obstacle or a platform, became a signal. It revealed hidden forces in the ocean. Brine rejection belongs to the same family of revelations. It takes something visible—the freezing of seawater—and turns it into a quantitative driver of the unseen.

To see why the salinity change is so potent, it helps to think in household terms. Imagine making ice cubes from saltwater in a tray. The first ice that forms will be less salty than the water you started with, so the leftover liquid becomes saltier. If you were to keep removing ice, the remaining brine would become increasingly concentrated, increasingly dense. In the ocean, the "tray" is the surface mixed layer: the top tens of meters that are continually stirred by wind and convection. When sea ice grows, it effectively removes freshwater from that layer while leaving salt behind. The mixed layer can therefore become denser than the water beneath, and once that happens, gravity takes over. The mixed layer overturns, sinking and drawing up slightly lighter water from below, which can then be cooled and salted in its turn.

This is convection in the everyday sense: denser fluid sinking beneath lighter fluid. But in the ocean, convection has an additional twist: it is trying to do its work in a rotating, stratified environment. *Stratification* means the ocean is layered by density, usually with lighter water above denser water. Stratification is stable, like a layered cocktail. Brine rejection is one of the few processes strong enough to punch through that stability, at least locally and seasonally. When it succeeds, it connects the surface to depth, injecting oxygen into waters that might otherwise remain isolated for centuries, and carrying surface properties—heat, carbon, chemical tracers—downward. This is one of the ways the ocean acquires memory: the surface leaves fingerprints that are stored in the deep.

There is also a human reason to care about this polar pump. Dense water formation influences the global distribution of heat. When dense water sinks at high latitudes, it is replaced by surface waters arriving from lower latitudes. Those incoming waters often carry heat. That heat release to the atmosphere can soften winters in nearby regions, while the exported heat deficit helps cool the planet as a whole. The details vary by basin, but the principle is robust: sinking in cold regions is connected, by the demands of mass conservation, to inflow along the surface. The ocean cannot send water down without bringing other water in.

The North Atlantic provides a classic example of this coupling. Parts of the Nordic Seas and the Labrador Sea are sites where surface waters can become dense enough to sink, influenced by winter cooling, sea ice processes, and the delicate balance between salty inflows and freshwater from precipitation, rivers, and melting ice. The exact locations and strengths change from year to year. But the broader idea has been stable enough to enter popular imagination: northern sinking helps pull warm surface currents northward. That image can be oversimplified, yet it captures something essential about the role of polar density production. The engine is not a single pipe, but it does have a throttle: alter the density of high-latitude waters, and you can alter the vigor of the deep overturning circulation that redistributes heat across the planet.

Sea ice participates in that throttle in two opposing ways. On one hand, making ice rejects brine and can promote sinking. On the other hand, a thick, persistent ice cover can insulate the ocean from the atmosphere, reducing heat loss and suppressing the very cooling that helps create dense water. A polynya, by keeping water exposed, resolves the paradox: it allows both intense heat loss and continuous brine rejection. That is why coastal polynyas can be disproportionately important. They turn the ocean surface into a conveyor of cold and salt, feeding dense water formation efficiently.

Even at the scale of a single winter storm, the process can feel almost alive. Strong winds tear at the ice, opening leads. Exposed water flashes its heat into the air, and frazil ice—a slurry of tiny crystals—forms in turbulent water. Those crystals collide, stick, and thicken into new ice. Salt is expelled into the surrounding water, increasing salinity. The surface layer becomes heavier and begins to sink in streaks and patches, while other areas are still forming ice. What looks like a frozen desert from afar can, at the interface between ice and water, be frenetic.

Tiny organisms inhabit that interface too. Sea ice is not sterile; it hosts microbial communities, algae, and small grazers that live in brine channels and at the ice underside. Brine rejection alters the salinity and temperature fields those organisms experience. It also influences nutrient supply. When dense water formation drives mixing and convection, nutrients from deeper layers can be brought closer to the surface, priming the system for blooms when light returns in spring. In polar regions, biology is often constrained not by nutrients but by light; still, the physical stirring sets the stage, determining when and where nutrients are available and how long they remain in the sunlit zone.

There is a deeper, more subtle connection between brine rejection and the ocean's chemical ledger. When surface waters sink, they carry dissolved gases with them. Oxygen is the most immediate: it is the breath of the deep sea, and it arrives at depth mainly through the sinking of surface waters. Carbon is more complicated, because it exists in multiple chemical forms and is influenced by biology, but the principle of transport is similar. Dense water formation can sequester surface carbon in the deep ocean for long periods. This is part of why polar processes matter for climate: they influence not only how heat is stored and moved but also how carbon is partitioned between ocean and atmosphere.

Brine rejection also sharpens the boundary between what the ocean remembers and what it forgets. The surface ocean is forgetful. Winds mix it; storms churn it; seasons reset it. A temperature anomaly at the surface can be erased in weeks. But when the surface is connected to depth by sinking, the anomaly can be exported below the reach of everyday weather. It becomes harder to remove, not because it is locked in a vault, but because the key—rapid mixing with the atmosphere—no longer fits. Dense water formation is, in that sense, a mechanism for converting short-lived surface events into long-lived subsurface signals.

This helps explain why changes in sea ice can have consequences that outlast the ice itself. Suppose sea ice formation decreases over a period of decades in a region that used to produce dense water efficiently. Less ice formation means less brine rejection and, often, less exposure of the ocean to intense winter cooling. The surface layer may then remain too light to sink as often or as deeply. In the short term, the change might look like a local shift in winter conditions. Over longer times, it can alter how much oxygen reaches certain depths, how heat is stored in the abyss, and how the deep ocean's properties evolve. Conversely, increased sea ice formation in some settings could enhance brine rejection and densification, though the balance depends on the competing insulation effect and on how winds, freshwater inputs, and ocean stratification respond.

The "engine" metaphor becomes almost literal when one considers the energy pathway. The atmosphere does work on the ocean through winds, but the deep circulation is powered largely by the conversion of potential energy: making dense water at the surface and letting it sink releases gravitational potential energy that can be dissipated in turbulence and used indirectly to maintain mixing. The ocean's stratification is continually being eroded by mixing and rebuilt by surface buoyancy forcing, and polar densification is a key part of that rebuilding. In that cycle, sea ice is not merely a passive boundary but an active participant in the energy budget.

One can stand on a winter shoreline in a high-latitude sea and feel how ordinary the ingredients are: cold air, salt water, wind. Nothing suggests that the same ingredients help maintain a circulation that links oceans and continents. Yet brine rejection turns these familiar elements into a precise, repeatable physical operation: a salt filter that runs at the freezing point, producing water heavy enough to sink. The water disappears below the surface, carrying with it a record of the conditions under which it formed—temperature, salinity, gases, and chemical traces. That record does not remain where it was written. It spreads, slowly and persistently, through the deep ocean, a cold signature stamped in winter and stored in darkness, powering motion by the simple insistence that heavy things fall.

7.2 The Thousand-Year Loop

A glass of water seems forgetful. Stir in a spoonful of dye and, within moments, the color spreads until the whole glass looks uniform. The ocean does not behave like that. Much of it is layered, reluctant to mix, and the deeper you go, the more the water carries the stubborn signature of where it last touched the air. That signature is not poetic; it is measurable. Oceanographers speak of *ventilation*: the act of bringing surface contact—and the gases, heat, and chemical fingerprints that come with it—into the ocean interior. A water mass is "ventilated" when it has recently been at the surface; it is "old" when it has been isolated from the atmosphere for a long time.

The age here is not the age of a molecule. Water molecules are constantly exchanged, split, and recombined; they do not carry passports. The age is more like the age of an idea in a rumor network: how long since this parcel of seawater last had a chance to exchange properties with the air above. In practice, that is inferred using tracers, substances whose concentrations change in time in known ways. Radiocarbon, for example, decays at a fixed rate, and it enters the ocean mainly through contact with the atmosphere. If deep water has less radiocarbon than surface water, it suggests the deep water has been out of touch with the atmosphere long enough for radioactive decay to matter. Other tracers, including anthropogenic chemicals released by humans in the twentieth century, act like time stamps of a different kind: if a deep sample contains no trace of them, it has likely not been at the surface since before those chemicals existed.

Those ages are not academic. They are the ocean's memory written in numbers. They tell you whether a patch of deep Pacific water last breathed the atmosphere hundreds of years ago or closer to a thousand, and they remind you that the ocean is not one big stirred pot. It is a set of connected reservoirs, with narrow gateways and slow exchange. The Global Overturning Circulation—often introduced through the popular image of a single "conveyor belt"—is the slow plumbing that links these reservoirs. It is not a rigid belt, and it does not carry each drop along a tidy path. It is better understood as a system of transformations: water is made dense in a few high-latitude regions, sent into the interior, modified by mixing and chemistry, and eventually returned toward the surface through widespread, gentle upwelling and currents that spread across basins.

The temptation to draw a single loop is understandable. Human minds like closed shapes. A loop promises comprehension: if you can trace it, you can know it. But the ocean's loop is not drawn in ink. It is drawn in probability. There are preferred pathways—deep currents do follow the contours of basins, and the rotation of Earth imposes strong constraints—yet the ocean also leaks sideways, mixes, and transforms water masses gradually. The overturning circulation is global because its consequences are global, not because it is a single pipeline.

One place where the loop is most tangible is the Atlantic. There, warm surface waters flow northward, releasing heat to the atmosphere, while colder, denser waters return southward at depth. This pattern is often called the Atlantic Meridional Overturning Circulation (AMOC), and it is one of the most climatically consequential expressions of overturning because it carries heat across the equator and toward Europe and the Arctic. The Atlantic is also relatively narrow and bounded, which makes its large-scale flow easier to detect than the sprawling, more complex exchanges of the Indo-Pacific.

Yet even in the Atlantic, the reality is more textured than the postcard diagram. Dense water formation does not occur as a single waterfall plunging straight to the bottom. It happens through a combination of winter convection, boundary currents, and the overflow of dense waters from marginal seas. Water transformed at the surface is guided along the western boundary of the basin by the rules of rotating fluids: in the Northern Hemisphere, deep currents tend to hug the western side of the ocean basin, a consequence of how Earth's rotation and the slope of the sea surface conspire to steer large-scale flows. Oceanographers have given this family of currents a memorable name: *deep western boundary currents*. They are narrow compared to the ocean's width, but they can carry a disproportionate fraction of the deep flow, like a hidden freight corridor.

A historical turning point in understanding these hidden corridors came with Henry Stommel's insistence, mid-twentieth century, that you could not explain the deep ocean with vague hand-waving about sinking and spreading. The ocean is a rotating fluid on a sphere with complicated boundaries; dynamics matters. Stommel and Arnold Arons (around 1960) developed an influential, deliberately simplified theory of how deep waters formed in high latitudes might spread through basins and return, not as an undirected seepage but through structured flows shaped by rotation. Their models were not the final word—no such thing exists in a moving ocean—but they shifted the conversation. The deep sea was no longer a stagnant backwater with occasional intrusions; it became a dynamical system with its own highways and rules.

The second turning point was less theoretical and more technological: the ability to measure. In the 1990s and 2000s, a network of moorings, floats, ship sections, and satellite observations began to turn the overturning circulation from an inferred idea into something observed, if still imperfectly. A celebrated example is the RAPID array at about 26.5°N in the Atlantic, a line of instruments designed to estimate the strength of the AMOC continuously. It has shown variability on seasonal to interannual timescales, reminding everyone that the overturning is not a steady conveyor running at constant speed. It pulses. It responds to winds, buoyancy forcing, and the chaotic variability of the atmosphere. The ocean's "thousand-year" character does not mean it is unchanging; it means that some of its consequences and adjustments play out on timescales longer than a human life.

To imagine the full global loop, it helps to step away from the Atlantic and think about what happens to water once it has sunk into the deep. The densest waters, formed around Antarctica, can fill the abyssal layers of multiple basins. Antarctic Bottom Water is often described as creeping northward along the seafloor, slipping through gaps in submarine ridges and flowing into the deepest plains of the Atlantic, Indian, and Pacific. "Creeping" is the right word if you picture speed, because deep currents are generally slow: centimeters per second rather than meters. But slow does not mean small. A current a few centimeters per second, hundreds of meters thick and hundreds of kilometers wide, transports an extraordinary mass of water.

As this deep water spreads, it is modified. It mixes with waters above and with waters of different origins. It gains heat very slowly from the surrounding ocean and from the seafloor. It can lose oxygen as microbes consume organic matter that rains down from the surface. It accumulates nutrients released by the decomposition of sinking biological material. In other words, it becomes chemically older even as it becomes physically altered. When oceanographers find that deep Pacific waters tend to be among the oldest in ventilation-age terms, that is partly because of geometry and pathways: the Pacific is large, and deep waters entering it must travel far and mix along the way before they return to the surface.

Returning to the surface is the part of the loop that is hardest to picture because it is not concentrated in one dramatic place. Deep water does not typically pop up as a single fountain in the middle of the ocean. Instead, it is lifted gradually by a combination of mixing and winds. Mixing provides a vertical leak: turbulence, often generated by internal waves breaking over rough topography and by interactions with boundaries, allows dense water to be transformed into lighter water at shallower depths. Winds, especially over the Southern Ocean, provide another crucial mechanism: they can drive surface divergence and upwelling, drawing deep water upward toward the light. The Southern Ocean matters here because it is the only latitude band where strong winds circle the globe unobstructed by continents. That continuity allows it to act as a global coupler, connecting basins through both surface and deep pathways.

If the loop were a single file of parcels, you could assign it a single travel time: a thousand years, say. The real ocean is more like a tangled set of routes with different travel times. Some water is ventilated and returned relatively quickly. Some becomes trapped in slow pathways and remains isolated for much longer. Modern modeling and tracer studies often suggest that typical deep-ocean ventilation timescales are on the order of a millennium, but with wide variation. In some simulations that examine how ventilation age changes across glacial cycles, global mean ages below one kilometer can sit in the neighborhood of several hundred to around a thousand years under different climate states, while the oldest deep waters—often in the Pacific—can exhibit much larger ages, especially during times when overturning patterns shift. These numbers are not mere curiosities; they are evidence that the deep ocean can store anomalies and release them long after the surface conditions that created them have changed.

A concrete way to feel the weight of these timescales is to consider a human-tagged tracer: chlorofluorocarbons (CFCs), once widely used in refrigeration and aerosols. They entered the atmosphere in large quantities in the mid-twentieth century, spread globally, and began dissolving into the ocean surface. Their production has since been curtailed by international agreement, but their legacy remains as a marker of "recent" ventilation. When oceanographers detect CFCs in a water mass, they can infer that this water has been at the surface since those compounds became common. The striking thing is where CFCs are *not* found: vast regions of the deep ocean remain free of these tracers, implying isolation from the atmosphere for longer than the industrial era. This is a rare kind of perspective: a chemical invented within living memory becomes a probe of circulation patterns that unfold over centuries.

That perspective matters for climate in two intertwined ways. First, the overturning circulation redistributes heat. When warm surface waters move poleward and release heat to the atmosphere, they shape regional climates. When deep waters carry cold and heat anomalies away from formation regions, they influence where heat is stored. Second, overturning circulation redistributes carbon and nutrients. A surface region that sends water to the deep can also send dissolved carbon, helping to sequester it away from the atmosphere for long periods. Conversely, where deep water returns upward, it can bring carbon-rich waters back toward the surface, affecting the exchange of CO_2 with the air.

Nutrients add an additional twist that connects physics to biology. Deep waters accumulate nutrients because organic matter produced at the surface sinks and is decomposed at depth. Where upwelling brings those nutrients back toward the surface, they can fertilize new biological production, including phytoplankton blooms that form the base of marine food webs. The overturning circulation therefore links the breath of the deep sea, the diet of surface ecosystems, and the carbon cycle, all on a timetable that is not synchronized with elections, quarterly reports, or even lifetimes.

The popular "conveyor belt" image can obscure a crucial point: the overturning is not simply a loop of water moving at a constant pace; it is a set of processes that can change in strength, structure, and location. Freshwater input from melting ice and increased precipitation can make surface waters less salty and therefore less dense, potentially inhibiting sinking in some regions. Changes in wind patterns can alter upwelling and the stirring of the surface ocean, influencing how effectively deep waters return. The system is responsive, and that responsiveness is exactly why the ocean can store memory: the response can be slow, lagged, and uneven across depths and basins.

This lag is not a flaw in understanding; it is a physical inevitability. The deep ocean is vast, and it is separated from the surface by stratification. If you change surface conditions today, you do not instantly change the abyss. Instead, you send a signal that must be carried downward through ventilation and mixing. Some signals penetrate quickly into the upper ocean; others take centuries to reach the depths where much of the ocean's volume resides. This is why deep-ocean changes can continue long after surface forcing stabilizes, and why the deep sea can act as a reservoir of past climates.

The sense of a "thousand-year" loop is therefore not a claim of clockwork regularity. It is a reminder of scale. The overturning circulation is one of the few physical processes on Earth that can, by itself, make the planet feel larger than a human life. It takes the seasonal act of densifying water at the poles and translates it into a global, multigenerational pattern of transport. It keeps chemical ledgers for centuries, storing the atmospheric past in deep water properties. And it creates a situation in which present-day actions can commit the ocean to future changes that will unfold slowly, in layers, long after the immediate weather has been forgotten.

To hold this in mind without losing the thread, it helps to return to the most concrete image of all: a bottle filled with layered liquids of different density. Tip the bottle slightly and you can set the layers sliding, but the densest liquid remains at the bottom unless something forces it upward. The ocean is that bottle, except the bottle is the size of a planet and the tilt comes from winds, freezing, heating, and the constant tug of gravity acting on slight differences in salt and temperature. In that bottle, the deep water formed in polar cold can travel across basins, acquire chemical history, and eventually be coaxed upward again, not in a dramatic surge but in a patient, persistent seep. The surface ocean changes its mind quickly; the deep ocean keeps its counsel, returning it only when the slow arithmetic of mixing and winds finally brings old water back into the light.

7.3 The Planetary Battery

The most misleading thing about global warming is how much of it is missing from the air.

Weather happens in the atmosphere, so the atmosphere feels like the natural place to look for the planet's extra heat. Thermometers sit in shade screens a couple of metres above the ground; headlines quote degrees of warming at the surface; heatwaves arrive as hot afternoons and sleepless nights. Yet the atmosphere is, in energetic terms, a thin rind. Most of the heating imbalance created by greenhouse gases is not parked in air at all. It is stored in water.

A useful word here is *inertia*, not in the everyday sense of laziness, but in the physical sense of resistance to change. Heat is a form of energy, and warming means adding energy. The question is how much energy it takes to raise temperature by a given amount. The technical phrase is *heat capacity*. Water's heat capacity is famously large: it takes far more energy to warm a litre of seawater by 1°C than it does to warm a litre of air by 1°C. Expressed as volumetric heat capacity (energy per cubic metre per degree), seawater is around $4.2 \times 10^6\,\mathrm{J\,m^{-3}\,K^{-1}}$, thousands of times larger than air. That single fact underwrites a great deal of climate experience that humans have learned to take for granted: coastal regions have milder seasons than continental interiors; fog can hover over cold currents on warm days; a warm ocean can keep feeding storms long after sunset.

Scale that intuition up from a bay to an entire planet and the ocean becomes something like a thermal flywheel. The atmosphere flickers from day to day; the ocean carries the long-term balance sheet. When extra energy accumulates in the Earth system, the ocean is where it mostly goes. Assessments summarized by the IPCC for the period from the early 1970s to the late 2010s put the ocean's share of the Earth system's energy gain at roughly nine-tenths, with the remainder divided among warming land, melting ice, and the atmosphere. The exact percentages shift with the accounting period and methods, but the broad picture is stable: the ocean dominates the energy inventory.

That dominance matters because it changes what "warming" means in practice. If most of the energy is going into the ocean, then the atmosphere can look temporarily less alarming than the underlying physics is. Surface air temperature can stall for a few years or even a decade while the planet is still accumulating heat. This is not a loophole in greenhouse warming; it is the ocean doing what a good heat reservoir does: absorbing energy without changing temperature very quickly at the surface, and then redistributing that energy in time and depth.

A good way to build intuition is to imagine a vertical column that contains the entire atmosphere above one square metre of Earth's surface, and next to it a vertical column of ocean above one square metre. The atmosphere is tall, but its density is low; the ocean is shallow by comparison, but dense. Because water stores so much energy per unit volume, only a few metres of ocean contain enough heat capacity to rival the whole air column above it. This is why sea-surface temperature anomalies can bully the atmosphere. A modest warming of the upper ocean can supply the energy that shapes humidity, cloud formation, and storm tracks; a modest cooling can pull heat from the air and stabilize it. The ocean is not just a passive sink. It is an active power source, releasing and absorbing heat on the timescales that matter for weather and climate.

The ocean's battery-like character is not only about capacity; it is also about *access*. The top layer of the ocean—the mixed layer—is continually stirred by winds and waves. It can be as thin as a few metres in calm, strongly stratified summer conditions or as thick as hundreds of metres in stormy winter conditions. Heat exchanged with the atmosphere is first stored in this mixed layer. Over time, some of that heat is transported sideways by currents and pushed downward by a combination of processes: wintertime convection in some regions, the slow heaving of density surfaces, and mixing driven by winds and internal waves. Once heat slips beneath the immediate reach of the atmosphere, it becomes harder to return quickly. It is not trapped forever, but it is sequestered on human timescales.

A historical anecdote shows how recently this view became measurable rather than rhetorical. For much of the twentieth century, the deep ocean was sampled mainly by ships. A research vessel would occupy stations, lower instruments on cables, take profiles, and move on. The data were precious but sparse, like a handful of weather stations trying to describe an entire continent. Then, around the turn of the twenty-first century, a quiet revolution began: autonomous profiling floats, the backbone of what is now widely known as the Argo program, started to populate the global ocean. These floats drift at depth, periodically dive and rise, and transmit temperature and salinity profiles by satellite. For the first time, the ocean's upper two kilometres could be monitored as a global, evolving field rather than a patchwork of expeditions.

That technological shift changed public debates in a surprisingly emotional way. In the early 2000s, surface air temperature rise appeared to slow relative to the rapid warming of the late 1990s. This was sometimes presented, incorrectly, as evidence that global warming had paused. Ocean observations helped reframe the question. A planet can continue to accumulate heat even if surface air temperature wiggles. Variability in winds and currents can push more heat into deeper layers for a time, leaving less in the surface and atmosphere. When enough of the ocean is observed, those pathways become visible rather than speculative: the energy is not missing; it is moving.

The global distribution of ocean heat uptake also connects to the circulation described earlier, but with a different emotional register. The overturning circulation is slow enough to feel abstract, yet it is one of the ways the battery is charged and discharged. In regions where surface waters are transformed and sent downward, heat can be exported into the interior. In regions of upwelling, older waters return toward the surface and can release heat back to the atmosphere. This is one reason the Southern Ocean looms so large in modern climate science. Its winds and circulation patterns provide an efficient gateway between the surface and the ocean interior, and changes there can alter how quickly heat is taken up and where it is stored.

Storing heat in the ocean does not mean storing it harmlessly. Batteries can power lights, but they can also power fires. Ocean warming has direct consequences that are felt at the coast and in the open sea. Thermal expansion contributes to sea-level rise: when seawater warms, it takes up more space, and that extra volume raises the height of the ocean even without adding a single drop of meltwater. Heat also stresses marine ecosystems. Corals, for example, live near their thermal limits; a sustained marine heatwave can trigger bleaching, an emergency response that can lead to mortality. The ocean's ability to absorb heat therefore buys the atmosphere time at the cost of loading the sea with energy that affects coastlines and food webs.

Because the ocean stores energy so effectively, it also commits the future to a kind of delayed response. Even if greenhouse gas concentrations stopped rising tomorrow, the ocean would continue to adjust. Heat already stored in the ocean would keep influencing sea level and the exchange of heat with the atmosphere. The planet has multiple lags, and the ocean is responsible for many of the longest ones. This is what makes "thermal inertia" more than a textbook phrase: it is a statement about irreversibility on timescales humans care about. Some changes can be slowed; some can be avoided; but some are already baked into the ocean's slow equilibration.

The accounting of where energy goes can feel abstract, so it helps to see it. The following figure is a stylized version of the Earth's energy accumulation since 1970, divided into ocean, land, ice melt, and atmosphere. The precise numbers depend on the dataset and methodology, but the point is qualitative and robust: the ocean's slice dominates the ledger.

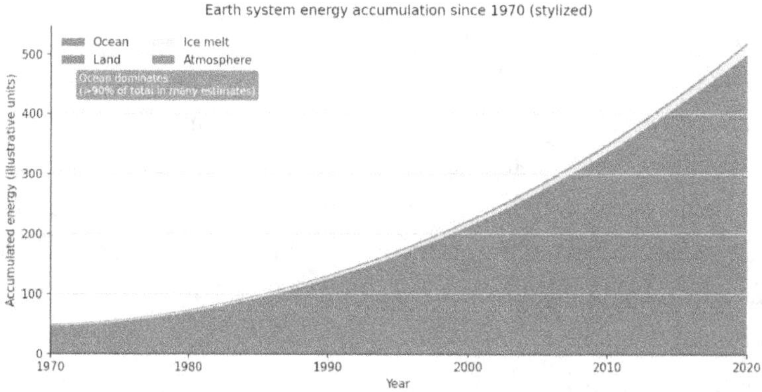

The battery metaphor becomes even sharper when you think about what the ocean is *made* of. Not just water, but moving water. A battery in a device is designed so charge can be stored and then released along a circuit. In the ocean, the "circuit" is circulation, and the release is not a tidy discharge but a redistribution. Heat stored in one region can be carried thousands of kilometres away. Heat stored near the surface can be mixed downward, reducing surface warming for a time but increasing subsurface warming. A warmer subsurface can then influence ice shelves from below, destabilizing them and changing freshwater input, which in turn can affect stratification and currents. The battery's wiring is tangled, and it runs through the lives of organisms as much as through physics.

One of the most personal ways this shows up is in humidity. Warm air can hold more water vapour, and water vapour is itself a greenhouse gas. The ocean supplies most of the atmosphere's water vapour. When the ocean warms, evaporation tends to increase, raising humidity and altering rainfall patterns. In many regions, this amplifies heat stress: the body's ability to cool by sweating depends on how readily sweat can evaporate, and that depends on humidity. A warmer ocean can therefore make a hot day feel worse even if the air temperature is the same, simply by changing the moisture content of the air.

The ocean's heat storage also complicates how people interpret cause and effect. A cold spell in winter can coexist with a warming planet because the atmosphere is noisy, while the ocean is the long-term accountant. A short-term slowdown in surface warming can occur while the ocean continues to absorb energy. In the language of risk, the ocean reduces the immediate signal-to-noise problem for surface temperature while increasing the commitment to future change through stored heat. It is a buffer and a liability at the same time.

There is a final twist that makes the battery metaphor slightly uncomfortable: unlike a household battery, the ocean did not consent to being charged. The additional energy it absorbs is not a service it provides on request; it is a consequence of a planetary imbalance. The ocean can take up heat because physics allows it, not because it is limitless. As the upper ocean warms, it becomes more stratified:

warm water is lighter, and lighter water on top tends to suppress mixing with colder, denser water below. Stronger stratification can slow the downward transport of heat in some regions, keeping more heat near the surface where it can influence weather extremes and ecosystems. The battery's capacity is vast, but the pathways into its interior can narrow or shift.

For the deep sea, the most sobering implication is that change can arrive quietly. The abyss is cold and stable by human standards, and many deep ecosystems are adapted to remarkably constant conditions. When heat penetrates to those depths, it does so slowly, but it does not do so gently. A fraction of a degree in a place that has been stable for centuries is a genuine upheaval. The ocean's memory is often described as slow; for organisms living on the edge of metabolic margins, slow does not mean safe.

A planet that hides most of its warming in the ocean is still warming. The energy is not gone; it is circulating, expanding water, lifting sea level, feeding humidity, and altering the background conditions against which storms and seasons play out. The air may be where heat is felt first, but the ocean is where heat is kept.

Chapter 8

Pulses, Poisons, and Prophecies

The ocean is not a static backdrop but a dynamic engine of rhythm and rupture, capable of shifting gears with devastating speed. We delve into the massive atmospheric seesaws that reorganize global weather patterns and the microscopic upheavals that turn fertile waters into toxic dead zones, revealing how delicate the balance between abundance and catastrophe truly is.

8.1 The Heartbeat of the Pacific

Stand on a pier in northern Peru on a good day and the ocean can feel like a generous machine. The water is cool for the latitude, the air often hazy, and the seabirds look slightly overconfident, as if gravity were a negotiable rule. Beneath the surface, that confidence is earned. Nutrients well up from depth, microscopic plants multiply in the sunlit layer, small fish gorge, bigger fish follow, and people build entire coastal economies on the reliability of that chain.

Then, every few years, the machine changes its settings.

The El Niño—Southern Oscillation (ENSO) is the most influential source of year-to-year climate variability on Earth. It is not a single storm or a single current. It is a coupled dance between ocean and atmosphere in the tropical Pacific: winds push water; water stores heat; heat changes where clouds form; clouds and pressure patterns reshape winds again. Like a heartbeat, it is rhythmic but not metronomic—strong in some decades, subtle in others, sometimes skipping, sometimes surging. And like a heartbeat, when it changes, everything downstream notices.

At the simplest level, ENSO toggles between states that scientists call *neutral*, *El Niño*, and *La Niña*. Those names can mislead. Neutral is not a do-nothing baseline; it is an active balance maintained by persistent winds and the ocean's response to them. El Niño and La Niña are not merely warmer or cooler water; they are reorganizations of the equatorial Pacific system—where the warmest water sits, how deep the nutrient-rich boundary layer lies, and where the atmosphere decides to build its towering thunderstorms.

Begin with the trades: the easterly trade winds that typically blow from the Americas toward Indonesia along the equator. Those winds do two crucial things. First, they push surface water westward. Second, they encourage *upwelling* in the east: as surface water is driven away from the South American coast and, more generally, away from the equator, colder water rises to replace it. Upwelling matters because deep water is often rich in dissolved nutrients such as nitrate and phosphate, accumulated from the slow breakdown of sinking organic matter. When those nutrients are lifted into the sunlit layer, phytoplankton respond quickly.

The westward push of the trades also piles warm surface water in the western Pacific. Sea level there can be tens of centimeters higher than in the eastern Pacific. The ocean does not like being tilted, but it can tolerate that tilt because the winds constantly reinforce it. The tilt has a hidden partner: the *thermocline*, the zone where temperature drops rapidly with depth. Under neutral conditions, the thermocline is deep in the west (warm water sits on top of warm water for a greater depth) and shallow in the east (warm surface water sits above cool water not far below). A shallow thermocline in the east makes upwelling especially effective: when water rises, it taps into cool, nutrient-bearing layers.

El Niño begins when this balance loosens. For reasons that can involve a burst of westerly winds, internal ocean waves, or the accumulated state of the coupled system, the trade winds weaken. When they weaken, the warm water that had been piled in the west *sloshes* back toward the central and eastern Pacific. The sea surface warms where it is usually cooler, and the thermocline flattens: deeper in the east than usual, shallower in the west than usual. Upwelling in the east still happens, but it draws from warmer, more nutrient-depleted layers. The biological engine along the eastern boundary starts to sputter.

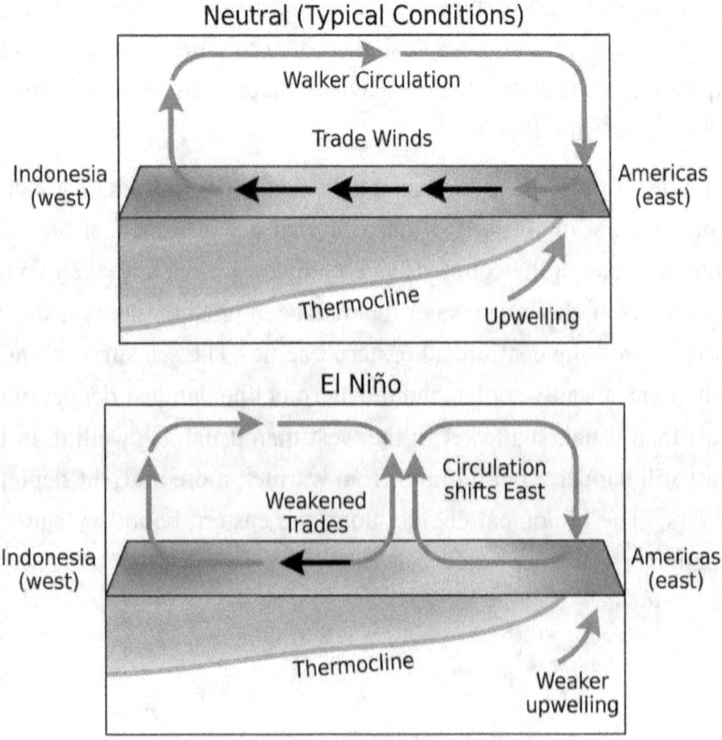

Cross-section along equator (Surface at top; Depth downward)

The oceanic reshuffle does not stay politely in the ocean. The atmosphere reacts to warm water because warm water evaporates easily, and evaporation feeds convection—the rising, condensing air that builds the tropics' tallest cloud towers. Where those thunderstorms form, surface pressure tends to drop; where air descends, pressure tends to rise. The tropical Pacific has a characteristic circulation called the *Walker circulation*: air rises over the warm western Pacific, flows eastward aloft, sinks over the cooler eastern Pacific, and returns westward near the surface as the trade winds. During El Niño, as warm water spreads east, the main region of tropical convection shifts east as well. The Walker circulation weakens and can reorganize. The pressure contrast across the basin changes; historically, that see-saw has been tracked using the Southern Oscillation Index, based on pressure differences between places like Tahiti and Darwin.

This is where the word *coupled* matters. Winds are not merely responding to sea-surface temperature; they are helping to create it. A classic reinforcement, described in the twentieth century by Jacob Bjerknes, ties these pieces together. Start with a small patch of warmer-than-normal water in the central or eastern Pacific. Warmer water encourages more convection overhead; convection alters pressure; weaker trade winds reduce the westward push of surface water and reduce upwelling of cool water; both effects warm the east further. The loop can strengthen an event from modest to major.

The biological consequences are immediate where people have learned to expect the ocean's generosity. The Peruvian and Chilean coasts sit beside one of the most productive upwelling systems on the planet. Under neutral conditions it is a nutrient pipeline: deep water rises, phytoplankton bloom, and small schooling fish—especially anchoveta—convert those microscopic meals into dense, fish-shaped energy that can be harvested at scale. In the early 1970s, Peru's anchoveta fishery was the largest in the world by tonnage. Then the system was hit by a powerful El Niño in 1972–73. Warm surface water spread east, the thermocline deepened off South America, and the nutrient supply to the surface weakened. Anchoveta, which thrive in cool, food-rich waters, suffered dramatic declines. Overfishing had already eroded resilience by pushing the population hard, but El Niño delivered the environmental punch: less plankton, altered habitat, and a food web suddenly running on a thinner budget. The collapse rippled outward—lost income, lost jobs, and a lesson that "maximum yield" calculated on a stable ocean is a fragile promise when the ocean has its own rhythms.

In a warm year, it is tempting to picture the ocean as simply hotter, but the key is often *stratification*: the layering of water by density. Warm water is lighter than cool water. When the surface warms and the thermocline deepens, the surface layer becomes more isolated from the nutrient reservoir beneath. In an upwelling system, the connection to depth is the difference between a green, productive sea and a clearer, hungrier one. The smallest organisms feel that change first. Phytoplankton are the ocean's primary producers, converting sunlight and dissolved carbon dioxide into organic matter. When nutrients become scarce, many phytoplankton cells shrink, growth slows, and the entire food web above them can lose power. Because phytoplankton also influence the biological carbon pump—the export of carbon to depth as organic particles sink—a shift in plankton productivity and community composition can alter not just fisheries but the ocean's role in the global carbon budget, even if the strongest, most immediate impacts are felt as missing fish.

ENSO's reach extends far beyond the equatorial Pacific because the atmosphere is an efficient messenger. When tropical convection shifts, it nudges the position and strength of jet streams and storm tracks through what climatologists call *teleconnections*. These are not magical long-distance controls; they are the atmosphere doing what it always does: adjusting to changes in where heat is released. The consequences can look like a string of coincidences until you learn to see the underlying pattern.

Consider the eastern Pacific during a strong El Niño: warmer-than-normal water, increased convection farther east, and altered upper-level winds. Those changes can make the subtropical jet over the Pacific more active and shift storm tracks. For the southwestern United States, this sometimes means wetter winters, while other regions can experience different outcomes. In other parts of the world, El Niño has been associated with drought risks, including in Australia and parts of Indonesia, as convection that normally dumps rain over the western Pacific and maritime continent relocates. The human meaning of this is not abstract: farmers timing planting, reservoirs managed for flood control and drinking water, public health shaped by heat and rainfall, and insurers quietly recalculating what "normal" should mean.

No two El Niño events are identical. Some peak in the eastern Pacific, others more in the central Pacific; some develop quickly, others build slowly; some trigger dramatic global impacts, others remain strangely modest. Part of the reason is that ENSO interacts with background conditions—the seasonal cycle, decadal patterns in the Pacific, and the chaotic weather noise that can either kick the system or damp it. Predictability exists, but it is limited. Forecast centers now track ENSO using satellites, moored buoys, and models that simulate the coupled ocean—atmosphere system. Yet even with impressive observing networks, forecasts often struggle with the "spring predictability barrier," a tendency for uncertainty to grow when the system passes through boreal spring, when the tropical Pacific's seasonal structure is in transition.

La Niña, the cool counterpart, is not merely "the opposite" in a symmetrical way, but it does reverse key features. Trade winds strengthen, warm water is piled even more strongly to the west, and upwelling in the east intensifies. The thermocline tilt steepens: shallow in the east, deep in the west. For eastern boundary ecosystems, this can mean a return of nutrient supply and productivity, though the precise ecological outcome depends on timing, local currents, and the life cycles of the organisms involved. Stronger upwelling can also mean colder coastal waters and, in some places, enhanced fog and different wind-driven hazards for mariners. Meanwhile, the shifted tropical convection and altered jet streams bring their own suite of teleconnected weather patterns elsewhere.

There is a subtlety here that matters for understanding why ENSO can feel like prophecy and betrayal at once. Many of ENSO's most dramatic impacts come not from the temperature anomaly itself, but from what that anomaly *does* to gradients—differences across space that power motion and mixing. A small change in the east—west temperature contrast can reorganize winds; a change in thermocline depth can switch the nutrient supply from generous to stingy; a shift in storm tracks can decide whether a winter rain arrives as replenishing water or as destructive floods. ENSO is a lesson in leverage: the planet's climate system has knobs that, when turned, move multiple levers at once.

The ocean, for its part, keeps records. During El Niño, heat that had been stored in the western Pacific is redistributed, changing not only sea surface temperatures but also the vertical structure of heat content. Because the ocean is Earth's largest heat reservoir, these shifts matter for global climate statistics. A strong El Niño can contribute to warmer global average surface temperatures for a year, while La Niña can do the opposite. But the more consequential story for many communities is local and biological: where the ocean breathes nutrients into the light, and where it holds them back.

Imagine being a fisher whose knowledge is intimately practical. You do not need to know the phrase "Bjerknes feedback" to notice that the water is warmer, the usual birds have thinned out, the fish are smaller or absent, and unfamiliar species appear. You might even welcome a new catch for a season, until you realize it is a symptom of displacement rather than abundance. ENSO can reshuffle marine neighborhoods: species track temperatures and prey; larvae drift differently in altered currents; predators follow the scent of opportunity. In some El Niño years, warmer waters can invite tropical species farther poleward; in others, the loss of upwelled productivity reduces the entire carrying capacity of the system, tightening the food web from its base.

This is why ENSO's "heartbeat" is not a gentle metaphor. A heartbeat sets the tempo for oxygen delivery, nutrient supply, and the capacity for work. When the tempo changes, the whole body must adjust. Along the Pacific rim, those adjustments are measured in different currencies: a failed anchoveta season, a costly flood, a drought that empties reservoirs, a coral reef stressed by unusual warmth, a public health system managing mosquito-borne disease after anomalous rains. Each is local. ENSO is the shared metronome behind them, beating across an ocean wide enough to hide its machinery from casual view.

On a map, the equatorial Pacific looks like open water between continents. In reality, it is a tight coupling between physics and biology, between wind and plankton, between the depth of a boundary layer and the price of fishmeal. The heartbeat is not audible, but it is measurable: in pressure differences across the basin, in the slope of an invisible thermocline, in the color of coastal water when nutrients surge—or when they do not.

8.2 When Blooms Go Bad

A plankton bloom can look like a miracle from the shoreline. The water turns jade or turquoise; sunlight catches on the surface as if the sea had decided to carry its own lanterns. In many places a spring bloom is the moment the ocean restocks itself, converting winter nutrients into new living tissue. Small grazers multiply, fish larvae find food at the right size, seabirds arrive as if they had been invited. A bloom is the ocean's way of saying: energy is back on the menu.

The same basic phenomenon can also close a beach, poison shellfish, kill marine mammals, and drain a coastal town's income in a single season.

Harmful algal blooms, usually shortened to HABs, are not one thing. They are a family resemblance: episodes in which algae (a loose term here that includes many kinds of microscopic photosynthetic organisms, and in fresh water often cyanobacteria) become harmful because of *toxins*, *overwhelming biomass*, or both. Some produce potent neurotoxins that move up food webs and into people. Some do not need toxins at all; they simply grow so dense that they shade out competitors, clog fish gills, or, most commonly, create the conditions for oxygen collapse when the bloom dies.

The most unsettling feature of HABs is that they are not always exotic disasters. Often they are the same ecological machinery that supports productive seas, pushed into a different gear by subtle changes in nutrients, temperature, and water movement. It is the difference between a well-timed bonfire and a house fire. The chemistry is similar; the context decides whether it warms you or burns you.

Start with what a bloom is in physical terms: a spike in the concentration of phytoplankton in the sunlit surface layer. Phytoplankton reproduce fast. Given light, a stable surface layer, and enough nutrients, some species can double in a day. They do not need to wait for a season to pass. They wait for conditions to line up.

Nutrients are a crucial part of that alignment. In the open ocean, phytoplankton are often limited by the supply of nitrogen, phosphorus, iron, or a mix of several. Coastal waters, by contrast, sit downstream of land, and land is where humans have become extraordinarily good at mobilizing nutrients. Synthetic nitrogen fertilizer, concentrated animal feeding operations, leaky sewage systems, urban lawns, and disturbed soils all send nitrogen and phosphorus into rivers and, eventually, into estuaries and nearshore seas. Those nutrients are not intrinsically evil; they are the same atoms that make proteins and DNA. The problem is dosage and timing. Ecosystems built for a steady trickle can be overwhelmed by a flood.

The result is *eutrophication*: nutrient enrichment that increases algal growth. Eutrophication is not identical to harmful algal blooms, but it loads the dice. When excess nutrients enter a water body, algae respond first because they can. That response can be benign, or even temporarily beneficial to some fisheries, until it tips into a kind of ecological crowding: too much biomass, too much decomposition, too little oxygen, too many opportunities for the wrong species.

The oxygen story deserves special attention because it is how a bloom turns into a suffocating event even without a single molecule of toxin. Oxygen in water is a limited resource. Cold water holds more dissolved oxygen than warm water, and mixing with the atmosphere is slow compared with breathing on land. Many coastal systems also stratify: warm, fresher water from rivers sits on top of cooler, saltier water, creating a density barrier that resists vertical mixing. When a bloom forms in the surface layer, it can pump oxygen into the water during the day via photosynthesis. At night, respiration dominates and oxygen declines. If the bloom is large, the day-night swings can be dramatic.

Then comes the inevitable: mortality. Cells die, get grazed, or clump and sink. Bacteria and other microbes feast on the organic matter, a process that consumes oxygen. If much of that organic matter ends up in bottom waters that are isolated from surface mixing, oxygen can fall to levels that stress animals and then to levels that can no longer support them. This is *hypoxia* (low oxygen). In the most extreme cases, it becomes *anoxia* (essentially no oxygen). Fish can flee if there is somewhere to flee to. Bottom-dwelling animals often cannot. Entire seafloors can become biological deserts for weeks or months, punctuated by carcasses and the few hardy organisms that tolerate low oxygen.

By the early twenty-first century, studies such as Diaz and Rosenberg (2008) described a dramatic global expansion of coastal dead zones, with many systems reporting worsening hypoxia since the 1960s. The pattern is a signature of modern nutrient management: we have become very good at fertilizing land and inadvertently fertilizing water, while also warming and stratifying many coastal seas.

One recurring example is the large hypoxic zone that forms in summer in the northern Gulf of Mexico, associated with nutrient loads delivered by the Mississippi and Atchafalaya rivers. The details vary year to year, but the story is consistent: spring and early summer deliver nutrients; phytoplankton grow; organic matter sinks; microbial decomposition consumes oxygen in stratified bottom waters; the low-oxygen zone expands across parts of the continental shelf. The U.S. Environmental Protection Agency's outreach in the mid-2020s framed this clearly as a linked problem of nutrient pollution, algal overgrowth, and oxygen depletion, with consequences for ecosystems and coastal economies. The map changes each year, but the mechanism is stubborn.

Toxins add a sharper edge. Many harmful blooms are harmful because certain species manufacture chemical compounds that interfere with nerves or liver function. Dinoflagellates, diatoms of the genus *Pseudo-nitzschia*, and other microscopic taxa can produce toxins that are harmless to the cells themselves but devastating to animals that consume them or concentrate them.

Shellfish are the classic conduit because they are efficient filters. A mussel or clam can clear liters of water per hour, capturing particles the way a well-designed sieve captures flour. During a bloom, that filter becomes a toxin concentrator. The shellfish may not look or taste different. The danger is invisible until it is measured. When people eat contaminated shellfish, the consequences can be severe: paralytic shellfish poisoning, amnesic shellfish poisoning, diarrhetic shellfish poisoning, and other syndromes whose names are blunt about what they do.

A widely remembered American example unfolded along the U.S. West Coast in 2015, when an unusually extensive bloom of *Pseudo-nitzschia* produced high levels of domoic acid, the toxin responsible for amnesic shellfish poisoning. Shellfish and crab fisheries faced closures and disruptions; sea lions and seabirds stranded in alarming numbers, disoriented or seizuring, after feeding on contaminated prey. The bloom was not simply "nature being nasty." It coincided with unusually warm ocean conditions in the northeast Pacific, including the marine heatwave nicknamed "the Blob," and with seasonal upwelling dynamics that influence nutrient supply and plankton communities. The event became a case study in how climatic anomalies can interact with coastal productivity to produce hazards that cross the boundary between ecology and public health.

Why do some blooms go toxic while others do not? There is no single trigger. Toxin production can be influenced by nutrient limitation, light, temperature, grazing pressure, and even microbial interactions, depending on the species. Yet a few themes recur.

One is nutrient *composition*, not just nutrient quantity. The ratio of nitrogen to phosphorus, the chemical form of nitrogen (nitrate, ammonium, urea and other organic forms), and pulses of nutrients after storms can favor certain taxa over others. Heisler and colleagues (2008) emphasized a consensus view that nutrient pollution promotes many harmful blooms, while also noting the complexity: not every HAB is nutrient-driven, and the same nutrient reduction can produce different outcomes in different systems. Anderson and Glibert's work in the early 2000s similarly highlighted that different harmful species exploit different nutrient niches. Some thrive on regenerated nutrients in coastal waters; some can use organic nitrogen forms that slip through standard monitoring categories. When a watershed delivers a cocktail rather than a single ingredient, it can select for organisms that are well adapted to that cocktail.

Another theme is water column stability. Many harmful bloom-formers do well when the surface layer is calm and stratified, because it keeps them in the light and reduces turbulent losses. Some dinoflagellates are not passive drifters; they can swim on the scale of meters per day, migrating up for light and down for nutrients, exploiting stratification like a daily elevator. Calm weather that feels pleasant at the beach can be the same calm that allows a bloom to intensify.

Temperature is a third theme, but it is rarely a simple "warm equals bad." Warming can speed up biological rates, expand the seasonal window for blooms, and strengthen stratification, all of which can favor certain HABs. But each species has its own comfort zone. Some harmful diatoms flourish in cool, nutrient-rich conditions; some cyanobacteria flourish in warm, stagnant waters. The more reliable statement is that warming increases the chances of mismatch: species appear earlier or persist longer, and the timing of nutrient pulses shifts with changing rainfall. Those timing shifts matter because plankton are opportunists. A storm that flushes nutrients into an estuary after a dry spell can act like striking a match near spilled fuel. The system can ignite quickly.

Freshwater HABs bring the story closer to home for people who never eat shellfish and live far from a coast. In lakes and reservoirs, cyanobacteria are often the main culprits. They are ancient organisms with modern advantages: some regulate buoyancy, some tolerate high light, some can fix atmospheric nitrogen under certain conditions, and many are well suited to warm, nutrient-rich, slow-moving water. Their toxins can affect the liver, nervous system, and skin. The public health pathway is direct: drinking water treatment, recreational exposure, pets that drink from shorelines. A late-summer lake that looks like pea soup can be less "algae" in the casual sense and more a dense microbial community whose chemistry can overwhelm municipal systems.

The economic and emotional toll of HABs is not only in medical outcomes. It is also in closures, uncertainty, and reputation. A coastal town can lose a season's worth of income because consumers hear "toxic algae" and stop ordering seafood, even if the local harvest is safe. Fish farms can suffer sudden mass mortalities when blooms deplete oxygen or irritate gills. Monitoring programs and rapid testing help, but the basic challenge is that harmful blooms are dynamic. A shoreline can be safe on Tuesday and risky on Thursday.

There is a grimly elegant cascade that turns nutrient enrichment into a dead zone, and it is so common that it can feel inevitable. Nutrients enter. Algae respond. Biomass accumulates. Light is blocked. The bloom collapses. Microbes consume the wreckage. Oxygen falls. The system simplifies. In reality, each arrow in that chain has detours. Grazers can intercept biomass and convert it into fecal pellets that sink faster. Wind can mix oxygen into deeper water and blunt hypoxia. A different phytoplankton community can build the same amount of biomass but decompose differently. Yet the cascade is a reliable narrative because it captures the direction in which excess nutrients push coastal waters: toward more organic matter and therefore more oxygen demand.

Once hypoxia becomes a recurring seasonal feature, it can alter ecosystems in ways that make recovery harder. Low oxygen stresses animals, reduces growth, and can change predator-prey interactions. Bottom communities shift toward species that tolerate low oxygen, often smaller-bodied and less diverse. Fish and shrimp distributions change, reshaping fishing effort and bycatch patterns. Even when oxygen returns, the memory remains in altered sediments and altered food webs.

The uncomfortable question is whether HABs are becoming more frequent or more severe. Part of the apparent increase is visibility: satellites now detect chlorophyll anomalies; monitoring programs track toxins; social media posts a photo of green water before an agency has finished its first sample. But there are also real drivers that plausibly increase risk: nutrient enrichment, warming, altered precipitation, and coastal development that changes circulation in estuaries and bays. The consensus view in the scientific literature does not claim a single global trajectory for all harmful blooms; it points instead to multiple regional increases and to mechanisms that make increases likely in many places, while leaving room for local exceptions.

The practical response begins upstream. Reducing nutrient loads can reduce the intensity of eutrophication and the oxygen demand that follows. Yet nutrient management is not a simple dial. Because nutrient *forms* matter, changing fertilizer practices, improving wastewater treatment, restoring wetlands that trap nutrients, and timing runoff reductions can all influence not only how much nutrient reaches the water but what kind arrives and when. In some systems, reducing one nutrient without the other can change community composition in unexpected ways. The goal is not merely starving the water; it is rebalancing it.

Downstream, monitoring and forecasting have become a quiet public service. Many regions run routine shellfish toxin monitoring and close harvests before illnesses occur. Forecasting systems combine satellite data, water sampling, and circulation models to estimate where blooms may drift and where toxins may accumulate. These tools do not eliminate uncertainty, but they turn surprise into risk management. A family deciding whether to harvest shellfish, a fishery manager deciding whether to open a season, and a hospital preparing for an exposure event all benefit from days of warning rather than none.

In a world of algorithms, there is an irony here: the smallest organisms can still outpace our ability to predict them. A bloom is the collective behavior of billions of cells responding to physics and chemistry that can change on the timescale of an afternoon breeze. The best forecasts are probabilistic, the way weather forecasts are, because the system itself is sensitive. That sensitivity is not a flaw of science; it is an honest description of living water.

There is also, buried in the hazard, a reminder of what productive waters are. The same nutrient-driven growth that can cause hypoxia is also the basis of coastal fisheries and the carbon transformations that make the sea a central player in planetary cycles. A bloom is not automatically a menace. It becomes one when the balance between growth, grazing, mixing, and decay tilts too far, when the water turns from a pantry into a trap. And when that happens, the consequences are not confined to the microscopic world. They land on dinner plates, hospital charts, and the calendars of people who make their living by reading the sea.

Chapter 9

Sentinels of the Deep

For most of history, our view of the ocean was limited to the view through a narrow soda straw, restricted by where ships could sail and when. We have since constructed a planetary-scale nervous system, a mesh of mechanical eyes and sensors that never sleep, revealing the breathing patterns of the global ocean. By synchronizing this autonomous fleet with rigorous human measurement, we are finally learning to read the water's memory and predict its volatile future.

9.1 The Autonomous Array

For most of the twentieth century, the ocean was studied the way remote deserts were once mapped: by rare crossings. A research ship would leave port with a plan, a stack of charts, and a kind of optimism that comes from knowing you only get one chance at each station. Crews lowered instruments over the side, waited as kilometers of cable played out, and then hauled the line back—hoping the sea stayed calm enough, hoping a squall did not arrive, hoping the winch behaved. Those measurements were precious, and they remain invaluable. But they were also snapshots taken under a moving ceiling of weather, politics, budgets, and ship time. Between expeditions, the ocean went on changing in the dark.

The shift that matters now is not merely from one instrument to another. It is from *visits* to *presence*. A modern observing system tries to behave less like a guest and more like a watchkeeper: always on station somewhere, always reporting, always accumulating a memory of what "normal" looks like so that change becomes visible. The Argo program—an international array of autonomous profiling floats—made that idea practical at global scale. Autonomous underwater gliders, slower and more deliberate, add a complementary talent: they can be steered into the places where the ocean's most consequential exchanges occur, along boundary currents and across fronts, repeating transects that ships can rarely afford to run often.

To understand why this is revolutionary, it helps to picture what a float does when no one is watching. A standard Argo float is not anchored. It is not remotely piloted minute by minute. It is set loose with a simple instruction set and enough energy to keep repeating it for years: sink, drift, dive, rise, talk, repeat. It is a machine designed to accept that the ocean will carry it where it will, and to make that wandering useful.

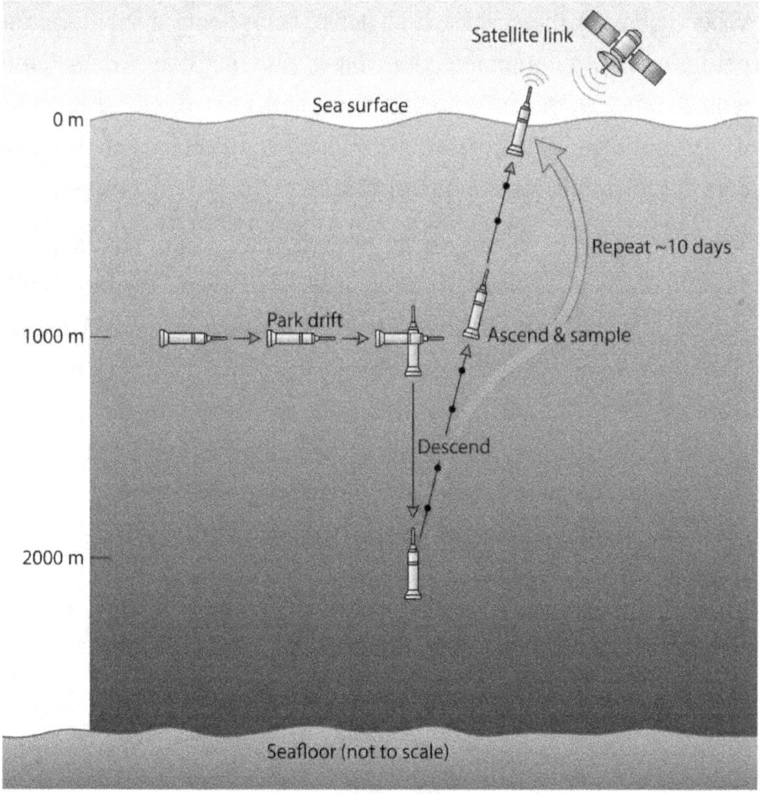

A float changes its buoyancy by shifting a small volume of oil between an internal reservoir and an external bladder. Expand the bladder and the float becomes slightly larger without gaining mass; its density drops, and it rises. Retract the bladder and it becomes denser; it sinks. The physics is homely—Archimedes' principle in an aluminum shell—but the implications are planetary. With that simple mechanism, a float can "park" at roughly 1000 meters depth, drifting with the currents that dominate the ocean's interior. On a schedule-often close to a ten-day rhythm-it sinks a bit deeper, commonly to about 2000 meters, then begins a measured ascent to the surface while recording a vertical profile of temperature and salinity.

When it breaks the surface it switches from being a submarine to being, briefly, a communications buoy, phoning home via satellite with the data it has collected. Then it sinks again, out of reach of waves and most predators of electronics, and becomes an invisible part of the circulation it is measuring.

The ten-day cycle sounds leisurely until you add up what happens when thousands of floats do it at once. The result is not a single perfect picture, but a continuously updating mosaic: the ocean interior as a field of evolving vertical slices, stitched together over time. Before Argo, much of the subsurface ocean—especially the Southern Ocean in winter, and the broad reaches of the South Pacific—was measured so sparsely that large-scale maps could be shaped as much by assumptions as by observations. Argo made the ocean's interior routine.

Routine is an odd word to celebrate. It suggests boredom, and in science boredom can be a warning sign. Yet in climate observation, routine is almost the whole point. The most dangerous changes are not those that announce themselves with drama, but those that accumulate quietly until they reshape the baseline: a slow freshening that alters stratification (the layering of lighter water over heavier water), a subtle warming stored below the surface, a creeping loss of oxygen that makes a habitat narrower year by year. Detecting such trends requires measurements that are not heroic; they must be *repeatable*. A float's schedule is its moral virtue: it does not get seasick, it does not get called back to port by a diplomatic crisis, and it does not decide that the weather is too ugly to work.

A concrete example of why this matters comes from one of the most important quantities in climate physics that almost nobody feels directly: ocean heat content. The ocean absorbs most of the extra heat trapped by greenhouse gases, and it does so mainly beneath the surface where the atmosphere cannot easily return it to space. If you only sample the ocean occasionally, you risk confusing a temporary rearrangement of heat—moved around by currents, storms, and seasons—with a long-term accumulation. Argo's steady profiling has allowed scientists to track how heat is distributed through the upper two kilometers of the global ocean with a confidence that was simply unattainable when measurements depended on ship tracks and sporadic campaigns. The story is not just that the ocean is warming; it is *where* the warming sits, how it migrates with circulation patterns, and how it changes the stability of the water column—stability that in turn shapes biology by controlling what nutrients can reach the sunlit surface.

The array did not appear fully formed. Like many international infrastructures, it grew out of a practical frustration shared across countries: the cost of ship time was rising, while the demand for global data was rising faster. Early prototypes of autonomous floats existed in different forms, but turning prototypes into a climate observing system required standardization. The floats had to sample in comparable ways, transmit in consistent formats, and pass through agreed quality checks so that a profile measured by a French float in the South Indian Ocean could be meaningfully combined with one measured by a Japanese float in the North Pacific. This was less glamorous than inventing a new sensor, but more decisive. A standardized data system turns countless individual instruments into something like a single, distributed telescope aimed inward at the sea.

That system also changed the rhythm of oceanography. An expedition produces a surge of data and then a long delay: samples must be processed, calibrations checked, papers written. Argo created a kind of scientific "present tense." Profiles arrive continually, and although the most careful versions of the data are refined later, the broad outline of the ocean's state is available quickly enough to feed operational forecasting and reanalysis systems—those hybrid products that blend models and observations to estimate what the ocean is doing even between measurement points. For people living along coasts, this matters because ocean forecasts influence search and rescue, fisheries management, and the prediction of marine heatwaves that can devastate kelp forests and coral reefs. Presence becomes service.

Gliders take the idea of presence and give it a different geometry. A profiling float is at the mercy of currents in the horizontal; its talent is the vertical profile and the global distribution. A glider, by contrast, is a slow swimmer. It also changes buoyancy to move up and down, but it converts that vertical motion into forward progress using wings. The speed is modest-often on the order of a walking pace-but the endurance can be remarkable, and the path can be planned. That makes gliders suited to repeated cross-sections across a current or around a shelf break, returning again and again to the same watery "gate" to count what passes through.

Imagine a boundary current—the swift, narrow flows that carry heat poleward along the edges of ocean basins. These currents can be fickle, meandering and shedding eddies, and they matter disproportionately for weather and climate because they control where the ocean releases heat to the atmosphere. Measuring them from ships is possible but expensive; measuring them continuously is harder still. A glider can patrol such a region for months, surfacing periodically to send data and receive new instructions, then diving again. It can carry sensors for temperature and salinity like a float, and often oxygen as well, sometimes optical sensors that hint at particles and plankton. In a world where the ocean's most consequential changes often occur along boundaries—between warm and cold water, between shelf and deep sea, between oxygen-rich and oxygen-poor layers—gliders provide a kind of persistence that ships rarely can.

The rise of autonomous observation, however, is not a story of machines replacing people. It is a story of machines creating new kinds of uncertainty that people must learn to manage. A shipboard scientist can watch an instrument as it is deployed, notice a kink in the cable, smell overheated electronics, decide to rerun a sample. A float, alone in the Southern Ocean, cannot ask for help. Sensors drift; biofouling (the slow colonization by microbes and tiny organisms) can alter readings; a pressure sensor can age; a conductivity cell can develop a subtle bias. Small biases matter because the ocean's signals are often small: a fraction of a degree over years, a slight salinity shift that changes density enough to influence circulation. Argo's success depends on careful quality control, including a delayed mode in which expert analysts compare float data to reference measurements and to neighboring observations, looking for the signature of sensor drift. The very existence of a delayed mode is a reminder that continuous data are not automatically trustworthy. They must be tended, even when no one is at sea.

There is also an ecological irony. The floats and gliders are built to be indifferent to weather, but the ocean is not indifferent to them. A float's vertical track crosses layers that differ in light, nutrient concentration, and biological activity. In the sunlit surface ocean, phytoplankton—the tiny photosynthetic organisms that form the base of marine food webs—respond rapidly to changes in mixing and stratification. A storm can deepen the surface mixed layer, pulling nutrients up from below; a period of calm can let the surface warm and stabilize, trapping plankton in bright water but starving them of nutrients. Physical profiles from Argo help interpret these biological swings, because they reveal the changing depth of the mixed layer and the position of the thermocline (the zone where temperature drops rapidly with depth). When floats are equipped with additional sensors—oxygen, nitrate, pH, chlorophyll fluorescence, particle backscatter—they begin to track not just the stage on which biology plays out, but aspects of the performance itself: photosynthesis and respiration, nutrient drawdown, the chemical fingerprints of carbon cycling. The ocean's metabolism becomes, in a limited but powerful way, observable at scale.

One can feel the difference between expedition science and continuous presence in the way anomalies are discovered. A ship-based program might notice an unexpected patch of low oxygen or unusually warm water only when the ship happens to pass through, and even then it might be hard to know whether that patch is exceptional or merely typical for that season. With an array, anomalies have a context. A marine heatwave, for instance, is not simply warm water; it is warm water relative to a baseline, persisting long enough to stress ecosystems. Continuous profiles help establish that baseline below the surface, where heat can hide and later re-emerge. In a similar way, oxygen trends are not just a single low reading; they are a long, slow shifting of a boundary that determines where fish can live and where microbes switch to different metabolic pathways, altering nutrient cycles.

A historical anecdote makes the contrast vivid. In the 1990s, oceanographers debated how much the ocean was warming and where the heat was going, but the observational record was uneven. Many regions were sampled along shipping routes or in summer; wintertime high-latitude observations were scarce, and the deep ocean was often a blank. Around the same period, a different observing technology—expendable bathythermographs, or XBTs—provided many temperature profiles from ships, but later analyses revealed biases tied to fall-rate assumptions and instrument design. The lesson was painful: even large datasets can mislead if their systematic errors are not understood. Argo was conceived in part as an answer to that vulnerability: a globally distributed system with standardized instruments, known sampling, and a deliberate framework for assessing uncertainty. It did not eliminate error, but it made error a quantity to be estimated rather than a shadow that haunted every conclusion.

There is, finally, a philosophical change embedded in these drifting machines. Traditional expedition oceanography often treated the ocean as something to be conquered by planning: you decide where to go, you go there, you sample, you return. The autonomous array accepts that the ocean is a moving medium that will always outmaneuver schedules. A float that was deployed to sample a broad region may be swept into an eddy field or carried across a front; a glider may fight a current for days to hold a transect. These are not failures. They are disclosures. The ocean's circulation is not just something to measure; it is something that *moves the measuring instrument*, writing its own signature into the data as trajectories and drift paths.

The planet now has a scattered, tireless set of witnesses in the water column, each one making small, disciplined observations and sending them upward through the thin skin of the sea to satellites and servers and, eventually, to the models that try to anticipate our climate's future. The romance of the research vessel has not disappeared, but it no longer carries the entire burden of knowing. The ocean is still vast, still undersampled in crucial ways, still full of processes that unfold faster than a ten-day cycle can catch. Yet the basic asymmetry has shifted. Instead of hoping to arrive at the right place at the right time, the observing system is increasingly already there, waiting, recording the quiet arithmetic of change.

9.2 The Human Benchmark

An autonomous float can report faithfully for years, but it cannot explain itself. When its oxygen sensor begins to drift, it does not feel doubt. When a membrane slowly ages or a tiny film of biology grows over an optical window, it does not raise a hand and ask whether the trend it is seeing is in the water or in the instrument. The ocean's new, continuous presence depends on an older, slower presence that is easy to overlook: people on ships, collecting the measurements that define what *true* means.

Ship-based oceanography looks old-fashioned only if one imagines its job is to compete with robots on coverage. It cannot. A research vessel is too expensive, too few, too bound to weather and scheduling. Its job is different. It provides a benchmark: measurements so carefully taken, so well calibrated, and so chemically complete that they can anchor everything else. Without that benchmark, a global array becomes a global rumor. With it, the array becomes evidence.

The benchmark starts with a deceptively simple act: lowering a package of instruments on a cable until it reaches the deep ocean and then hauling it back. The core device is called a CTD, short for conductivity, temperature, and depth. Conductivity tells you salinity because salt makes seawater conduct electricity; temperature is straightforward; and depth is inferred from pressure, since the ocean presses harder the deeper you go. A CTD cast produces a vertical profile of the ocean's basic physical state. Mounted on the same frame are bottles—Niskin bottles, usually arranged in a ring—that can be closed at chosen depths to capture water samples. Those samples are where ship science becomes irreplaceable. They allow measurements that are too delicate, too slow, or too demanding of reagents and calibration gases to be done by a sensor drifting alone.

One of the first surprises for newcomers on oceanographic cruises is how much of the work is about *trust*. The ocean is hard on instruments. Salt corrodes, pressure crushes, and motion loosens connectors. Even a perfect sensor will not stay perfect forever. Shipboard teams treat every measurement as something to be earned: by calibrating thermometers against standards, by checking salinity against laboratory analyses of discrete bottle samples, by running blanks and replicates, by comparing oxygen sensors to chemical titrations. The aim is not only accuracy today, but traceability: a chain of procedures that lets someone, years later, understand why the number should be believed.

The need for this becomes clearer when thinking about trends. The most important signals of climate change in the ocean are often subtle compared with everyday variability. A storm can mix surface waters and change temperature and salinity dramatically over a day; an El Niño event can rearrange heat across an entire basin over months. Against that noisy background, the long-term drift of, say, salinity in the subtropical gyres or oxygen in the tropical thermocline may be small in any single year. The difference between a real trend and a sensor artifact can be thinner than a human hair. If a float's salinity reading is biased by a tiny amount, the derived density field shifts, and that can distort estimates of ocean circulation and sea level. If an oxygen sensor drifts downward, the ocean can appear to be losing breath faster than it truly is. Conversely, a drift upward can mask a genuine decline. Benchmarks are how one prevents the ocean from being misdiagnosed.

This is why programs such as GO-SHIP, the Global Ocean Ship-based Hydrographic Investigations Program, matter so much. GO-SHIP organizes repeat hydrographic sections across entire ocean basins, revisiting the same transects roughly every decade with tightly specified measurement practices. The goal is not speed but fidelity: full-depth profiles, coast to coast, including not only temperature and salinity but also a suite of chemical and biological tracers. These transects are a kind of audit. They tell you, with high confidence, how the ocean's inventories of heat, freshwater, carbon, nutrients, and oxygen are changing, and they provide reference-quality datasets against which autonomous and satellite observations can be checked.

A concrete example shows the chain of dependence. Consider salinity. An Argo float estimates salinity from conductivity measured by a small cell through which seawater flows. Over time, that cell can develop a bias because of fouling, aging, or subtle changes in the sensor. Argo's data system anticipates this. Float profiles undergo real-time quality checks, but they also enter a delayed-mode process in which expert analysts compare the float's data to a reference field constructed from shipboard CTD measurements and from other verified observations. Ship CTDs are used as a trusted baseline precisely because they are carefully calibrated and because their deployment allows direct bottle samples that can confirm salinity with laboratory precision. The delayed-mode system does not merely flag obviously bad data; it can detect slow drifts that would otherwise slip into climate analyses unnoticed. The autonomy of floats, paradoxically, makes human expertise more valuable: when you cannot hold the instrument in your hand, you must know the ocean well enough to recognize when the instrument is telling an implausible story.

Some of the most valuable benchmark measurements are of substances that exist in seawater in tiny amounts but have outsized interpretive power. Dissolved inorganic carbon, total alkalinity, nutrients such as nitrate and phosphate, and trace gases like chlorofluorocarbons (CFCs) are not just numbers; they are clues about processes. Carbon and alkalinity reveal how much anthropogenic (human-added) CO_2 the ocean has absorbed and how that absorption is altering acidity. Nutrients reveal the potential for biological production and the history of water masses, since nutrients are consumed at the surface by phytoplankton and regenerated at depth by decomposition. CFCs, introduced to the atmosphere by industrial use in the twentieth century and later regulated, act as time stamps for ventilation: they mark how recently a parcel of water was at the surface.

These are not measurements a typical autonomous float can yet do at the same accuracy, or with the same range, or with the same interpretive richness. A ship can take a water sample, preserve it, run it through a calibrated instrument in a shipboard lab, and compare it to reference materials. That reference material culture is one of the quiet triumphs of modern oceanography: the ability to say that an alkalinity measurement made in one ocean, by one team, in one year, is comparable to a measurement made by another team decades later. Without that, the global carbon budget—how much carbon the ocean is taking up, where, and how quickly it is changing—becomes an argument about methodology rather than a statement about the planet.

A historical anecdote captures what is at stake. When Charles David Keeling began precise measurements of atmospheric CO_2 in the late 1950s, he did something radical: he insisted on precision and continuity. The resulting curve from Mauna Loa, climbing year after year with a seasonal breathing pattern superimposed, became one of the most iconic records in Earth science. Oceanographers, watching that curve rise, faced a harder challenge. The ocean does not offer a convenient mountaintop. Its carbon system is buffered and dynamic, and the signal of anthropogenic CO_2 is spread through chemical forms and across depth. Detecting the ocean's uptake required measurements precise enough to resolve small changes in a large background. Time-series stations such as the Hawaii Ocean Time-series (HOT) north of Oahu and the Bermuda Atlantic Time-series Study (BATS) in the Sargasso Sea were established in the late 1980s with a similar philosophy to Keeling's: go back to the same place again and again, measure carefully, and accept that the power lies in patience.

Those stations are not glamorous in the way a one-off expedition to an ice edge can be glamorous. Their value is that they turn the ocean into a place with a memory. At HOT, repeated measurements have documented rising dissolved inorganic carbon and a gradual decline in pH (a measure of acidity) in the surface ocean, alongside shifts in biology and nutrient cycling that respond to climate variability. At BATS, long-term data have revealed how changes in winter mixing influence nutrient supply to the surface and therefore the productivity of phytoplankton. These are the kinds of records that allow a researcher to say, with confidence, that a decade of change is not merely a decade of unusual weather. They also serve as calibration points for autonomous sensors: if a float's pH sensor passes near a time-series site, comparisons become possible that would be difficult anywhere else.

Fixed-point observatories extend this logic. OceanSITES, a global network of moored stations, keeps watch at specific locations with instruments mounted on lines from the surface to depth, measuring temperature, salinity, currents, carbon system variables, oxygen, and meteorology. A mooring does something neither ships nor drifting floats can do well: it resolves fast variability at one place. It captures the passage of storms, the daily cycle of heating and cooling, the pulsing arrival of internal waves, the sudden injection of oxygen-poor water onto a shelf, the brief bloom of plankton that can happen and vanish between float profiles. Those high-frequency records are essential for understanding mechanisms, and mechanisms matter because they determine whether a trend will persist.

There is a temptation to talk about the ocean's "metabolism" as though it were a single thing. In reality it is a layered accounting system. Near the surface, phytoplankton use sunlight to convert dissolved CO_2 into organic matter, releasing oxygen. Below, microbes and animals consume that organic matter, using oxygen and releasing CO_2 and nutrients. This vertical separation—production near the surface, remineralization at depth—is what makes the biological carbon pump: a process that moves carbon from the atmosphere into the deep ocean for years to centuries. Autonomous platforms can now sense parts of this pump indirectly through oxygen, nitrate, fluorescence, and particle backscatter. But verifying those signals, and interpreting them in terms of carbon budgets, often requires bottle samples and shipboard measurements: particulate carbon collected on filters, pigments analyzed in the lab, rates of photosynthesis inferred from incubations, and a suite of chemical constraints that can disentangle biology from physics. A float might tell you that oxygen has changed at 500 meters; a ship can help tell you whether that change reflects altered circulation, changed respiration, or a sensor that has quietly lost its calibration.

There is also the matter of the deep ocean, where change is slow and the stakes are large. The deep sea holds an enormous reservoir of heat and carbon. Small changes there can matter greatly for sea level, because warming expands water, and for the long-term fate of anthropogenic carbon, because the deep ocean is where carbon can be stored away from the atmosphere. Deep Argo aims to extend profiling floats to the full depth, but deep measurements are technically demanding, and the benchmark role of ships becomes even sharper. Full-depth CTD casts, with carefully calibrated sensors and water samples, remain the standard against which deep autonomous data must be judged. If the question is whether the abyss is warming by a few thousandths of a degree per year, the difference between insight and illusion can be a calibration coefficient.

Ships offer another kind of benchmark that is less often discussed:

the benchmark of *context.* An oceanographer on deck can see the sea state, the clouds, the color of the water. They can note a slick of algae, an unusual odor, a swarm of gelatinous plankton, a change in the behavior of seabirds. These observations are not always quantitative, but they can guide hypotheses and interpret anomalies. They also remind us that sensors are not observing "the ocean" in an abstract sense; they are observing a lived, messy environment where biology can clog inlets, where bubbles from breaking waves can confuse optical instruments, where a thin lens of rainwater can make the surface layer behave differently from what a deeper profile might suggest. A benchmark is partly a number and partly the wisdom of how that number can be fooled.

None of this diminishes the autonomous array. It makes it more impressive, because it reveals the scaffolding that holds it up. A global system needs both breadth and depth in the human sense: breadth of coverage from robots, depth of certainty from ships and long time series. When those pieces work together, the ocean becomes legible in a new way. A float profile that looks odd is no longer discarded as noise or accepted on faith; it becomes a prompt. Is an eddy transporting unusual water into a region? Has oxygen been drawn down by a burst of respiration following a plankton bloom? Is a sensor drifting? The answer often requires a ship, not because the ship is faster, but because it can bring the full laboratory to sea and because it can measure the ocean with a kind of ceremony: careful calibration, discrete samples, repeated checks, and the hard-won humility that comes from knowing how easily water can lie.

On a calm night in the subtropics, a CTD package breaks the surface after hours in the deep, dripping seawater that has been out of contact with the atmosphere since before the last time a child learned to read. The deck lights catch the metal frame and the closed bottles. Someone checks the pressure readout, someone logs the station number, someone draws the first sample with a practiced hand. It is not dramatic, but it is a form of stewardship. The global ocean is too large to be watched only by people, and too consequential to be watched only by machines.

9.3 The Predictive Horizon

The ocean is not only a place where things happen; it is a place where the consequences of our decisions are stored. Heat slips beneath the surface and stays there. Carbon dissolves, reacts, and becomes part of a chemical system that can hold it for decades, centuries, longer. If the ocean is a buffer, it is also a ledger: it keeps accounts on behalf of the atmosphere, and it does not offer receipts that are easy to read.

Observation alone cannot turn that ledger into foresight. A map of temperature and salinity profiles, however dense, does not by itself tell you how the ocean will move next month, or how much heat it will take up next decade, or how the chemistry that controls acidity will evolve as emissions continue. For that, scientists do something that feels, at first glance, like alchemy: they combine measurements with equations, turning a scattered set of real numbers into a continuous, physically consistent estimate of the ocean's state, and then use that estimate to project forward. This is where the sentinels of the deep become part of a predictive machine.

The core idea is simple to say and hard to do: the ocean follows the laws of physics and chemistry, but we do not know its exact condition everywhere. We measure it imperfectly, and we model it imperfectly. Data assimilation is the craft of making these imperfections cooperate. It is the method by which observations are blended with a numerical model so that the resulting ocean estimate fits the data while still obeying the model's dynamics. In weather forecasting this is now an everyday term; in oceanography it has been transformative, because the ocean is both less observed and more memory-laden. A storm in the atmosphere can blow through in days; a temperature anomaly in the ocean interior can persist for months or years, carried and slowly mixed. That persistence is both a promise and a trap: it creates predictability, but it also means that errors, once introduced, can linger.

A numerical ocean model is not a video game ocean painted convincingly on a screen. It is a set of equations that represent the conservation of mass, momentum, heat, and salt, solved on a three-dimensional grid. The model must also represent processes that occur at scales smaller than the grid can resolve: turbulence, internal waves, mixing along density surfaces, and the exchange of heat and freshwater with the atmosphere. These unresolved processes are not mere details; they control how rapidly heat and carbon are moved downward and how quickly the ocean can respond to changes at the surface. Because they cannot be explicitly simulated, they are *parameterized*: represented by simplified formulas with tunable coefficients. Every parameterization is a bet about how the ocean behaves when you cannot see the details.

Observations keep those bets honest. A model can generate a plausible ocean even if it is wrong in subtle ways. Without observational constraints, it may drift toward a state that is internally consistent but not the real world. Satellites offer a wide-angle view: sea surface temperature, sea level from altimetry, ocean color as a proxy for surface chlorophyll. Autonomous floats provide vertical profiles through the upper ocean, and sometimes biogeochemical variables as well. Ships provide the benchmark measurements needed to ensure that the observational streams themselves remain accurate. Data assimilation is the point where these strands are tied together: the surface viewed from space, the interior sampled by drifting robots, and the chemical truth supplied by careful sampling.

One way to picture assimilation is to imagine trying to reconstruct a symphony from a handful of microphones placed around a concert hall. Each microphone hears something real, but each is biased: one is too close to the violins, another muffles the bass, a third crackles occasionally. A model of the symphony, built from the musical score and the physics of sound in the hall, can fill in what the microphones miss. Yet if the model is wrong about the hall acoustics, it can produce a beautiful reconstruction that is still not the performance that occurred. The best estimate comes from letting the microphones correct the model while letting the model interpret the microphones.

In practice, assimilation systems vary in how they do this. Some adjust the model state directly, nudging temperature and salinity fields toward observed values in a statistically optimal way. Others treat uncertain inputs as controls to be adjusted: winds, heat fluxes, mixing parameters. A prominent example of the latter approach is state estimation, where the goal is not simply to make the model match observations at each time, but to produce a trajectory that is dynamically consistent over long periods, closing budgets of heat and salt without allowing unphysical sources or sinks. The Estimating the Circulation and Climate of the Ocean project (ECCO) is often cited in this context; its value lies in treating the ocean as a constrained system whose history must make sense as a whole, rather than as a sequence of snapshots.

This insistence on budget closure matters for questions that society increasingly asks. How much heat has the ocean taken up over the last thirty years? Where has it gone? How much carbon is the ocean absorbing today, and is that uptake changing? These are global accounting questions, not merely local descriptions. They are also questions where the answer depends on the integrity of the physical system. If a model can match sea surface temperature while quietly creating or destroying heat in the interior through numerical artifacts, it might look accurate while giving a misleading ocean heat content trend. A state estimate that is forced to conserve heat and salt while fitting observations offers a sturdier foundation for diagnosis.

The predictive horizon is the boundary where this diagnosis becomes a forecast: a statement about what is likely to happen, not only what is happening. Ocean forecasts are already operational in many contexts, from days to months, and they are increasingly coupled to atmosphere models. Coupled models matter because the ocean and atmosphere continually exchange heat, moisture, and momentum. The ocean does not merely respond to weather; it shapes it, especially in regions where sea surface temperature anomalies influence atmospheric circulation. For seasonal forecasting, this coupling is familiar in the context of El Niño and La Niña, where changes in the tropical Pacific can reshape rainfall patterns worldwide. For longer timescales, coupling is central to the climate question: whether the ocean will continue to take up a large fraction of our excess heat and carbon, or whether feedbacks will alter that buffering role.

A concrete example of coupling that feels personal is marine heatwaves. When a region of the ocean surface stays unusually warm for weeks to months, ecosystems can be pushed past thresholds. Kelp forests can collapse, coral reefs can bleach, fish distributions can shift. These events are not only local tragedies; they are also experiments run by nature on the coupled system. A persistent atmospheric high-pressure pattern can reduce winds, suppressing the mixing that normally brings cooler water up. The ocean's stratification can intensify, making it even harder for surface heat to escape downward. Biological responses can then feed back on the physics: changes in plankton communities can alter how sunlight is absorbed in the surface layer. Forecasting such events requires an observing system that detects the buildup of subsurface heat and a model that can represent the relevant feedbacks.

The ocean's capacity to absorb carbon is an even more subtle kind of coupling. When atmospheric CO_2 rises, more CO_2 dissolves into the ocean surface. But dissolution is only the start. Dissolved CO_2 participates in a set of reactions known as the carbonate system, forming carbonic acid, bicarbonate, and carbonate ions. The balance among these forms determines pH and the availability of carbonate ions that many organisms use to build shells and skeletons. At the same time, biology alters carbon chemistry through photosynthesis and respiration: phytoplankton draw down dissolved CO_2 in the sunlit layer and export some carbon downward as organic particles, while remineralization returns CO_2 at depth. Circulation then moves carbon-rich water around the globe and, eventually, can bring it back to the surface. The ocean's carbon sink is therefore not a single mechanism; it is an interplay of solubility, chemistry, biology, and circulation.

Earth System Models attempt to represent this interplay. They extend physical climate models by including biogeochemical cycles: carbon, nutrients, oxygen, and sometimes explicit ecosystems. These models are used in coordinated international efforts such as CMIP (the Coupled Model Intercomparison Project), which provide the backbone for many climate assessments. But biogeochemistry amplifies the challenge of predictability. Physical variables like temperature and salinity are hard enough; chemical variables add new sources of uncertainty and new opportunities for bias. If a model's mixing is too vigorous, it can ventilate the interior too quickly, affecting oxygen and carbon storage. If its phytoplankton growth response to light and nutrients is too strong or too weak, it can misrepresent the biological carbon pump. If it mishandles alkalinity, it can misrepresent acidification trends and the buffering capacity of seawater.

Observations again act as a referee. Biogeochemical Argo floats can now measure oxygen, nitrate, pH, and optical proxies related to chlorophyll and particles, offering a broad, if imperfect, view of the ocean's metabolism. But the credibility of these measurements depends on calibration and cross-checking, and their interpretive value depends on being embedded in a physically consistent estimate of currents and mixing. Assimilating biogeochemical data is harder than assimilating temperature and salinity because the relationships are more nonlinear and the error characteristics less well understood. A pH sensor has different failure modes than a thermometer. A nitrate estimate depends on optical absorption that can be influenced by other substances. Yet the direction is clear: the observing system is becoming chemical, and the predictive system must become chemical too.

A historical anecdote from a neighboring domain is useful because it reveals how forecasting can stumble even when models look sophisticated. In 2012, the Arctic summer sea ice extent plunged to a record low. Many forecasts made that season underestimated the decline. The reasons were not a single flaw but a tangle: imperfect initialization, gaps in observations, limitations in representing key processes such as melt ponds and ice dynamics, and the difficulty of translating a forecast into a user-relevant product. The lesson carries over to the ocean. Better equations alone do not guarantee better predictions. Forecast skill depends on whether the model represents the right processes, whether the observations constrain the right variables, whether the assimilation system properly weights uncertainties, and whether the forecast is evaluated in ways that matter for decisions.

This is where the idea of a "predictive horizon" becomes less like a line on a chart and more like a shifting coastline. Predictability is not uniform. Some aspects of the ocean are predictable farther ahead because they have long memory. Subsurface heat content in certain regions can evolve slowly enough to allow seasonal-to-interannual forecasts with useful skill. Large-scale patterns of circulation can persist and provide a degree of inertia. But other aspects are inherently harder. Eddies, storms, and small-scale turbulence can scramble local conditions quickly. Even a quantity that seems integrative, like ocean heat content in a basin, can become hard to predict on decadal timescales if model biases in winds, stratification, or mixing systematically steer the system wrong. Ensemble forecasting—running many simulations with slightly different initial conditions or model formulations—is one way to cope, not by eliminating uncertainty but by mapping it.

The growing interest in machine learning adds another twist. Data-driven methods can emulate expensive model components, accelerate assimilation, or discover patterns that help with bias correction. In oceanography, there is genuine excitement about hybrid approaches that respect physics while using machine learning to fill gaps: for example, learning subgrid mixing tendencies from high-resolution simulations, or using neural networks to flag suspicious sensor data. Yet a purely data-driven predictor has a fundamental vulnerability in a changing climate: it learns from the past, and the future may not be statistically familiar. Physical consistency matters because it provides a kind of guardrail. When a model obeys conservation laws, it cannot conjure heat from nowhere, even if it would help match a noisy dataset. When a machine learning system is trained without such constraints, it can produce answers that look plausible and still violate the ocean's basic accounting. The challenge is not to choose between physics and data, but to build systems where data sharpen physics and physics disciplines data.

All of these technical efforts serve a human-scale question: how long will the ocean continue to protect us from the full force of our emissions? The ocean has absorbed most of the excess heat and a substantial fraction of the carbon dioxide produced by human activity. That buffering has bought time, but it is not free. Heat uptake contributes to sea level rise through thermal expansion and by influencing ice sheets. Carbon uptake drives ocean acidification, shifting the chemical environment in which marine organisms evolved. The same processes that allow the ocean to store heat and carbon—stratification, circulation pathways, the balance of mixing and ventilation—can change as the climate warms. Warming tends to strengthen stratification in many regions, making the surface ocean more resistant to mixing. That can slow the downward transport of heat and carbon in some contexts, while also reducing the supply of nutrients to surface waters, potentially altering biological productivity and the biological carbon pump. Meanwhile, circulation changes can redistribute where uptake occurs, concentrating impacts in particular regions.

The predictive system, at its best, turns these complexities into conditional statements that can guide choices: *if* emissions follow a given trajectory, *if* certain feedbacks behave within a plausible range, *then* the ocean's sink for heat and carbon is likely to evolve in these ways, with these uncertainties. That is not prophecy. It is a disciplined way of not being surprised.

On any given day, the flow of information is almost mundane. A float surfaces in heavy seas, transmits a packet of numbers, and sinks again. A satellite passes overhead, measuring the height of the sea surface to within centimeters, revealing the broad pressure patterns that steer currents. A mooring records a sudden burst of mixing as a storm passes. A ship, somewhere else, calibrates a sensor against a bottle sample and quietly improves the trustworthiness of thousands of profiles. In a data center, an assimilation system takes these disparate signals and adjusts a model state so that it becomes more like the ocean. The outcome is a set of fields—temperature, salinity, currents, oxygen, carbon variables—that are not observed everywhere but are constrained everywhere, a kind of continuously updated best guess.

The ocean will always keep some secrets. It is too large, too turbulent, too full of life to be fully captured by instruments and equations. But when the observing system and the predictive system are joined, the ocean becomes less like a black box and more like a living, legible component of the climate system: one whose changing capacity to absorb heat and carbon can be tracked, tested, and—within limits that matter—anticipated.

www.ingramcontent.com/pod-product-compliance
Lightning Source LLC
LaVergne TN
LVHW022205200926
848481LV00006B/1854
9798896652687